GRASSHOPPER DREAMING

GRASSHOPPER
D · R · E · A · M · I · N · G

Reflections
on Killing
and Loving

Jeffrey A. Lockwood

SKINNER HOUSE BOOKS
Boston

Printed in Canada.

Cover design: Kathryn Sky-Peck.

Cover illustration: Grasshopper by Tamas Galambos. Copyright (c) Tamas Galambos/SuperStock, Inc. Private Collection/Bridgeman Art Library, London/SuperStock.

Text design: Terry Bain

ISBN 1-55896-431-2

Library of Congress Cataloging-in-Publication Data

Lockwood, Jeffrey Alan, 1960-
    Grasshopper dreaming : reflections on killing and loving /
    Jeffrey A. Lockwood.
        p. cm.
    ISBN 1-55896-431-2 (alk. paper)
    1. Grasshoppers. 2. Grasshoppers—Ecology. 3. Insect pests—
    Control—Environmental aspects. I. Title
    SB945.G65 L64 2002
    179'.1—dc21   2002022681

05 04 03 02
9 8 7 6 5 4 3 2 1

To Nan, Erin, and Ethan.

Whether or not you understand, you believe.

Table of Contents

I do not know whether I was then a man dreaming I was
    a butterfly,
or whether I am now a butterfly dreaming I am a man.
                    —Chuang-tzu

Prologue

"IT'S MINE, DON'T TOUCH IT!" Mrs. Hanson ordered as the goldfish flopped desperately at her feet. In seventh grade, my big issues were whether Norma Blocker would go to the basketball game with me and if Retin-A would cure acne. I didn't need my social studies teacher presenting me with a life-and-death decision right after lunch, but the fish was now making only the occasional, pathetic squirm in front of the class. "Geez," Eric Palmer winced under his breath and rose to rescue the rapidly asphyxiating fish. Eric was one of the cool kids who could get away with defying a teacher, especially when a weird classroom experiment was evidently under way. Even for adolescents accustomed to being shocked by hip teachers in the early seventies, netting a fish and dumping in on the floor was a bit much. "Sit down," Mrs. Hanson commanded, glaring at Eric, and he dropped back into his seat. Mouths agape, we silently implored her to save the fish, which now lay still except for a weak flaring of its gills. "It's mine, and I can decide to kill it if I want," she declared. We slumped in our seats, resigned, confused, stunned. The fish's wide, un-blinking eye seemed to plead for help. Finally, she bent down, gently took the fish in her cupped hand, and slipped it back into the water, where it recovered quickly, darting around the bowl. As gentle as a mother's goodnight to a child, she whispered, "What does it mean to own a living thing?"

Good for Nothing

*The vigorous, the healthy,
and the happy survive and multiply.*
—Charles Darwin

INTRODUCING MYSELF at social gatherings as an entomologist is almost sure to generate interesting conversations. Everybody has stories to share and questions to ask about their encounters with insects. I haven't kept count, but the most common question I hear at parties goes something like, "I know we shouldn't kill them all, but really, what are they good for?" *Them* refers to the particular insect that is the topic of discussion. After a moment, most people often suggest their own answer: "I suppose birds eat them." But somehow this doesn't seem satisfactory, and they want me to explain the purpose of mosquitoes, miller moths, or grasshoppers.

I admire our increasing awareness that all beings are part of an interconnected whole and that when a strand of the web is broken, there are often systemwide effects. All of that is true, but it suggests, however implicitly, that the purpose of this web somehow involves us humans. The problem is that nature doesn't exist for us, ecosystems don't care about us, animals don't generally love us, and the universe doesn't really need us. Nearly two thousand years ago the Roman emperor and Stoic Marcus Aurelius counseled that it was important to "desire every one of your actions to be right in your own judgment, and remember two things: Your actions are significant, but the circumstances in which they take place have no significance." This paradox is compelling: Each life is of infinite value to itself and of no importance to the universe. To ask what a life, human or insect, is "good for" presumes that value lies in utility, that worth is not intrinsic.

I know grasshoppers. I've dedicated my professional life to their study. Over the past fifteen years, I have employed many methods for learning *about* grasshoppers. Only recently have I begun to consider what I might learn *from* them. Science provides innumerable tools for learning *about* life, but ultimately one must turn to other ways of knowing to discover how we might learn *from* it. Fortunately, one can interpret a single experience from multiple perspectives; "good science" need not preclude intuitive insight or transcendent understanding. An observation can provide information, foster knowledge, or evoke wisdom, depending on what the observer brings to the encounter. So it is that the grasshoppers have taught me, among other things, the nature and value of nothing.

When I was hired in 1986 as an assistant professor at the University of Wyoming, specializing in grasshopper biology and management, I was a competent entomologist. My only firsthand encounters with these insects, however, were as a child in New Mexico. My parents built a house on the outskirts of Albuquerque, and their landscaping provided the only green food the grasshoppers saw in the summer. This afforded bountiful opportunities for catching grasshoppers, which I housed in various containers or fed to the black widow spiders that lined the wall of the backyard, but these were hardly the experiences on which to build a scientific career.

So I began my university research by spending my first summer studying grasshoppers on the short-grass prairie just north of Fort Collins, Colorado. The site was ideal because it supported an abundance of grasshoppers (ten to fifteen per square yard), provided a wide diversity of species (thirty different kinds of grasshoppers), and was convenient (only an hour from the campus at Laramie where, at an elevation of 7,200 feet, summers are too short to generate many grasshoppers on the surrounding grasslands). I would simply park at a campground along Highway 287, climb over the barbed-wire fence, cross the weedy pasture to the north, clamber up a rocky slope, weave between the mountain mahogany shrubs, and emerge in a grassy field that covered a few hundred acres and contained a few million grasshoppers.

The ancient fellow who owned the land lived in a dilapidated set of buildings you could get to only via a bridge surmounted by two wooden tracks about eighteen inches wide. The challenge was to line up your truck perfectly on the dirt road leading to the ravine and then, unflinching, cross the bridge in a straight line. I suspect the old man didn't make the crossing often. In fact, except for the times I renewed his permission to access the land, I never saw him. I'm pretty sure I had his permission, but given his lack of hearing and teeth, I was never sure that he knew what I meant or that I knew what he said.

I spent hundreds of hours from June to September sitting on the prairie with a video camera recording grasshopper behavior. I decided to focus on just one species that year. The first rule of science is to simplify the problem, to isolate that which you seek to understand. I chose *Aulocara elliotti*, the bigheaded grasshopper. Although this is a serious pest of the rangeland, my interest was rather more pedestrian. *A. elliotti* was abundant and the size of a pencil stub, so it could be identified from a discrete distance. It has, as one might infer, an abnormally large head, along with a white X on its back and blue hind tibiae, or limbs.

Previous field studies on insect behavior had taught me that the greatest virtue of my summer's work would be patience. Remaining motionless to capture the behavior of the grasshoppers in an undisturbed state became increasingly difficult as summer progressed. The chill of dew-dampened mornings gave way ever more quickly to the searing heat of midday. The grasses set seeds, which took the form of variously modified darts that worked their way into socks, creases, and bootlaces. The sweat bees showed no gratitude for their feast, delivering burning stings whenever trapped between clothing and flesh. The muscular tension of holding the camera on my shoulder was creating a permanent crick in my neck. I took breaks periodically, but real relaxation came only when I became fully engaged in my filming, when I lost all sense of time and discomfort by total absorption in the life of grasshoppers. I didn't analyze the ten-foot shelf of videotapes until later that fall, but even in the summer I knew full well what grasshoppers did most of the time: nothing. Absolutely nothing.

The results of the summer of 1986 became my first published paper on grasshoppers. To be honest, only by a careful selection of particularly intriguing behaviors was I able to find enough activity in the videotapes to generate scientifically interesting conclusions. I determined, for example, that this species engages in territorial behavior, or at least aggressive intolerance of other individuals. Despite my focus on the times when the grasshoppers were "doing" something, for forty-three minutes out of every hour, they were not doing anything. They just sat there on the ground or hung in the vegetation. I called this "resting," presuming they were saving energy for the real demands of life. Other biologists have ascribed such immobility to thermoregulation: If they are sitting in the sun, they are really engaged in warming themselves, while if they are hunkering in the shade, they are cooling themselves. There are some occasions when these interpretations are valid, as when they perch on top of grasses and turn perpendicular to the sun's rays in the early morning. But I suspect that most of the time when grasshoppers appear not to be doing anything, they aren't clandestinely engaged in pursuing other goals—they are simply doing nothing.

From the perspectives of ecology and evolution, spending hours engaged in doing nothing is difficult to explain. After all, these grasshoppers suffer a daily mortality rate of about 2 percent, meaning that only about one-third of those that hatch in the spring will survive to reproduce as adults. Imagine how your workplace would change if one out of every fifty employees died every day. If your doctor told you that you had a 2 percent chance of dying each day, that would mean you could count on even odds of being dead by the end of the month. Under these circumstances, an organism should be desperately engaged in securing resources and assuring its biological success—eating and mating—especially when the essential ingredients are presumably in short supply.

Ecology and evolution are implicitly grounded in the structure of human economics. These explanatory systems presume that the dynamics of life arise because essential resources are limited, thereby necessitating brutal competition. Economists seem fond of developing models that further assume that perfectly informed agents act with perfect rationality to acquire these inadequate resources.

That is, of course, a silly assumption in the case of humans, who are often misinformed and profoundly irrational. But insects ought to be well informed by their finely tuned senses, and in the presumed absence of self-awareness and individual volition, grasshoppers should be largely driven by the cold, calculating logic of natural selection. Humans might invest according to their horoscopes, but insects ought to manifest behaviors that arise from simple cause-and-effect optimization of their fitness.

However, grasshoppers defy the economics that use either energy or genes as the currency of life. Grasshoppers are incredibly blasé about reproducing or feeding. Sex appears to be an activity of modest interest, at best. Courtship and mating occupy a small proportion of their days; most of their encounters seem to be more antagonistic than romantic. In fact, reproductive behavior was so rare that I excluded it from my analysis and titled my paper "Nonsexual interactions in *Aulocara elliotti.*" If we consider that grasshoppers often reach population densities of thirty, forty, and up to one hundred per square yard, surely they ought to be competing fiercely for their share of the food. But in my summer of behavioral recording, the grasshoppers spent only about three minutes out of every hour eating, despite the impending famine. There was no tragedy of the commons, no gluttonous devouring of a dwindling larder, no headlong race for each to extract the most food from the pantry. Indeed, in my many years of working with these insects in the field, I have encountered only three or four situations in which it seemed they had eaten themselves into an absolute shortage (a shortage of nutritious food may occur well before a field is literally stripped of all vegetation), and in these cases they simply walked or flew less than a day's journey to greener pastures.

What are the grasshoppers up to? If we humans were short of resources, we would surely battle for our share. We'd scurry about attempting to vanquish competitors, hoard supplies, mate feverishly, and, well, do much of what we seem to do in the modern world. But grasshoppers aren't humans. It is not even clear that they are operating under an economy of shortages, and if they are, there is scant evidence that they are behaving to ensure a competitive advantage. Why should they? If science aspires to objectivity, why is it appropriate to ascribe to other beings the values we use to explain or rationalize our actions? In a great subjective leap, we presume that competition for limited resources is the leitmotif of all living beings because this theme defines our own interactions in and with the world.

The fact is that grasshoppers spend most of their time doing nothing (unless you count digesting, breathing, and being incidentally warmed or cooled). Our struggle to understand their languor arises from our approaching these creatures with the same question with which we approach one another: "What do you do?" It is as if we can define all worth in terms of what someone or something does. This assumes that value is instrumentally derived—things have worth in terms of what they do for us. Relationships are critically important to defining life, but they are not the sole measure of our lives.

If we were to reconstruct our scientific understanding in the context of intrinsic value (the notion that something can have worth in and of itself), a rather different interpretation of animal behavior, ecology, and evolution would emerge. If we seek to reveal the inherent

worth and dignity of life—the intrinsic reality funda-
mental to Alfred North Whitehead's metaphysics, the
inner being essential to Teilhard de Chardin's understanding of existence, the self that was the core of Ralph
Waldo Emerson's meaning of life—then it is not sur-
prising that a grasshopper might spend a couple of hours
just sitting. I am reminded that Thich Nhat Hanh, the
Buddhist priest, suggested that when people are hurry-
ing about and shouting, "Don't just sit there, do something!" the crisis might be more effectively addressed if
a quiet voice admonished us, "Don't do something, just
sit there." Maybe grasshoppers would make good Bud-
dhists. They certainly defy the Protestant work ethic (a
failing immortalized in the children's tale "The Ant and
the Grasshopper") and the cultural values that underlie
scientific inquiry. Sometimes I wonder why we call ourselves "human beings," when we spend very little time
"being." Perhaps we ought to call ourselves "human do-
ings" and reserve the notion of "beings" for the other
creatures.

The do-nothing grasshoppers have taught me that science is very effective at assessing and understanding sub-
stance and activity. In terms of the interdependent web of
all existence, science excels at analyzing and controlling

the strands, but has little to say about the spaces. And a web is mostly empty space. There are ancient methods for exploring the space between the strands, but these methods are generally viewed as being not just different from but inimical to science. However, the emptiness is so essential to being that science must acknowledge its existence. Unable to manifest humility or reverence, we conquer the void by dint of language and faith. Naming is a powerful means of asserting control, and science has developed a rich assortment of terms to establish intellectual dominion over the elusive and unknowable. The mysteries that emerge between the strands are labeled as variation, noise, error, and chance.

This tactic would be more plausible if science had a test for randomness, but none exists. An appeal to randomness is a faithful prayer to the unseen. We can tell you what it isn't (randomness is the absence of identifiable pattern, the modern version of *terra incognita*), but we cannot assert what it is. Chaos theory demonstrates that sometimes randomness is constrained, shaped into a cloud of realized events by a so-called strange attractor. The origin and nature of these forces that sculpt order from formless chance are themselves a complete mystery. But what science cannot fathom, nature still manages to exploit. At every scale, creative order arises from putative disorder. In evolution, random mutation plays a pivotal role; in quantum physics, probability waves lie at the heart of existence; in cosmology, nothingness gave rise to the universe.

A resting grasshopper is akin to randomness; it manifests a behavior that fails to fit any identifiable purpose or pattern that we expect to see. It is, as far as we can tell, doing nothing and persists in this state of meaningless existence for prolonged periods of time. To code the behavior of these insects, I designated "resting" as 0. When analyzing data, we differentiate between "missing data" and "true zero." Missing data are just that: empty, information-free spaces because we didn't look (or lost the data). A true zero means that we looked but didn't see anything. This system, however, presumes that when we don't see anything, there really is nothing there. Categorizing resting behavior as a true zero created the illusion that I knew there was nothing other than an immobile, impassive, nonfeeding, nonmating, noncompetitive, uncommunicative organism devoid of biological meaning. What it really meant was that I didn't know what the grasshopper was doing, or whatever it was doing, it didn't fit any of my expectations of what a grasshopper ought to be doing. The latter interpretation is certainly suggested by some subsequent work on feeding behavior.

Grasshopper feeding presented a bit of a problem. The amount of forage that the ecologists claimed these insects consumed could not be reconciled with the amount of time the behaviorists had documented as devoted to feeding. This wasn't a major biological controversy, but it was an intriguing riddle. The solution was incredibly simple, being a matter of changing expectations and assumptions about what other life forms ought to be doing. The dogmatic description of a typical grasshopper's day involved its basking in the early morning, feeding at mid to late morning, sheltering to avoid the midday heat, feeding in the late afternoon, and then resting throughout the night. Nobody had actually spent any real time trying to watch grasshoppers at night. After all, they were obviously active during daylight, and staggering around the prairie at three in the morning seemed pointless, if not masochistic. But by capturing grasshoppers and examining their crop (stomach) contents over several twenty-four-hour periods, we discovered that on warm summer nights, these insects are happily munching away. In fact, some species decidedly prefer midnight snacking, which makes good sense given the risks of exposing themselves to predators while clambering around in the grass during the day. By partitioning their mealtimes throughout the day and night, a rich diversity of grasshoppers in a community effectively manages to feed continuously. Our discovery did not shake the foundations of science, but it did demonstrate how science can become a subjective projection of our lives, wants, and needs onto other organisms.

A good colleague recently told me that he had failed to replicate my findings of nocturnal feeding. He had taken several individuals of a single species of grasshopper from the field during the day, caged them in observation tanks, offered them prefabricated wafers of compressed grass throughout the night, and observed their feeding. They didn't eat the wafers, so he concluded that my "night feeding phenomenon" was an interesting but spurious result. I maintain that if you take an animal from its complex habitat, place it in a completely alien setting, and offer it a single, artificial food, then you probably can't say very much about what that animal and its community of related species are doing in the intact habitat. In fact, my experience with taking grasshoppers from the field and caging them generates a fairly consistent behavior within a very short period: They die.

First, dismissing unexpected results as spurious is a way to evade reality. In a spectacular metaphysical feat, that which science cannot explain ceases to exist. The life sciences are very good at induction and rather weak at deduction. We can predict the pattern or extract the generality, but we cannot explain the particular or account for the exception. For some matters this is fine, but the limit is obvious when we realize that each of our lives—and the lives of other beings—is ultimately a singular occurrence.

Second, science may be adept at developing and applying analytical methods, but you cannot see what you do not look for. Sometimes you don't even recognize you are looking through the wrong end of the telescope. Isolating elements of complex systems for scientific study is a defensible and useful tactic, but it requires that we ignore vast numbers of relationships. This approach generates valuable information and suggests plausible mechanisms, but it does not reveal ultimate explanations or assure wisdom.

And so, in answering the polite and honest question "What is a grasshopper good for?" the ecologist in me wants to discuss the role of this creature in nutrient cycling, and the evolutionist in me wants to explain that it is good at replicating itself. But I have come to understand that these are ends that we impose and values that emerge only by induction; the grasshopper is unaware of our goals and statistical extrapolations. We might as well as ask ourselves what our children are good for: Do we love them because they are efficient omnivores, effective competitors, successful phenotypes, genetic successors? These qualities give the right answers to the wrong question. The reason we value our children is not because of what they do, but because of who they are. That's why as a spiritual scientist, my answer is that a grasshopper isn't good for anything. Its presence is of no significance—an ultimate zero. Its value is in being a grasshopper, nothing more. The grasshopper just is. And that is enough.

Shall I Lead?

O body swayed to music, O brightening glance,
How can we know the dancer from the dance?
—William Yeats

ENTOMOLOGISTS CAN'T DANCE. Although this assertion might be substantiated by attending the mixer during a meeting of the Entomological Society of America, my claim extends to the world of modern agriculture. On this cultural dance floor, pest managers defy the rhythm of nature. We march to our own cadence, while the insects tango, samba, and jitterbug. Like General Patton taking the floor with Ginger Rogers, the asynchrony is often comic and sometimes tragic.

I've dedicated my professional life to learning to dance with grasshoppers on the rangelands of Wyoming and watching others either seek or defy the complex rhythms of these creatures. My first lessons were learned through pragmatism, where I either followed their lead or failed in my work. Initially, it was as if some practical joker painted numbered footsteps onto the studio floor for two different dances. The basic operations in ecological or management studies of rangeland grasshoppers are counting them and catching them.

With a reasonably abundant infestation, counting the grasshoppers within a fixed sample area is like determining the number of head bangers in a mosh pit. As you move through the grassland, pandemonium ensues, with grasshoppers pelting your legs and torso. The bedlam overwhelms any attempt to count them. An untrained observer typically overestimates population densities by at least twofold. But each grasshopper takes only a single jump, so the chaos is a matter of our perception. Through disciplined practice you learn to visualize and focus on an area of one square foot, and it becomes possible to see and count individuals. By synchronizing our scale with that of the grasshoppers, frenzied scattering transforms into single arching leaps.

Catching grasshoppers takes two forms—mass collecting and individual stalking—and each requires its own dance. Mass collecting is accomplished by sweeping a heavy muslin net attached to a stout handle through the grass and brush. With a hundred strokes of the net through an infested field, a new worker will slash a considerable swath of vegetation, snaring a couple dozen grasshoppers. On the other hand, my research associate will sweep the net in graceful arcs, gathering a couple hundred individuals. With a decade of experience, he has learned that the secret is to follow the grasshoppers' lead. You can't do the Charleston if your partner is waltzing. Small, immature nymphs require that the net be swept in a low, slow arc to intercept them in mid leap. But sweeping too slowly means that they will land before the net arrives, and sweeping too fast passes the net over them before they leap. Conversely, adult grasshoppers flush like tiny quail, so if the net is not swung high and fast, the yield will be meager. The choreography is further modified by temperature (when it is cool your partner doesn't move quickly, so the dance is slowed), time of day (in the early morning your partner basks high in the canopy, so the dance is elevated), and species (some of your partners relish explosive leaps, others prefer to dip). The standard approach is to take separate sets of low, slow sweeps and high, fast sweeps—and the two samples reveal entirely different troops performing on the prairie.

To collect a specimen of *Trimerotropis*, a wary genus that blends seamlessly into its surroundings, one must move with grace or fail. Like a villager pantomiming the hunt, the collector walks briskly with darting eyes. He freezes, gaze locked on a distant goal. Now with each fluid step, he sinks toward the earth until one knee is gently planted. A lithesome reach into the grass . . . a moment's pause . . . a strike . . . and the quarry is deftly pinned to the ground. This is an illusionist's dance, a performance for an insect audience. Grasshoppers know that an object is approaching if its size increases. By shrinking in stature while advancing, the insect perceives that the collector is stationary—until it is too late.

A dancer must accommodate both the music and a partner. Applied entomology often adapts to neither. In my field, we have long sought the music—the external cues—that underlie the annual fluctuations in grasshopper populations. To no avail, we have tried matching their dynamics to the rumba of rainfall, the tango of temperature, and even the swings of sunspots. So we tried time-lagged models, as if the grasshoppers were a few beats behind the orchestra. They weren't. Recent efforts show that grasshopper population dynamics may be self-generated, a dance driven by an internal tune, albeit con-

strained by their partners. Indeed, through badly con-
ceived control programs, we have cut in on the predators
and parasites. Their slow recovery from insecticide ap-
plications leaves the grasshopper population uncon-
strained. And like the girl in the fairy tale "The Magic
Shoes," we find ourselves pulled into a dance we cannot
escape.

A remarkable asynchrony has developed between the
grasshoppers and the U.S. Department of Agriculture.
Major outbreaks arise every seven to ten years, which is
just enough time for the feds to dismantle their infra-
structure from the previous program under the mistaken
belief that the dance is over. Just as they're rolling out
the rug and setting up the chairs, the band strikes up a
tune. As the grasshoppers sweep over the prairies, the
federal and state agencies rush to get in step. I take some
perverse pleasure in listening to my African colleagues
describe their efforts to synchronize the arrival of spray
planes and locust swarms, with the former inevitably
two steps behind. It reminds me of an old film in which
some poor bloke was taking a crash course in ballroom
dancing. Hopelessly lagging a step behind his gorgeous
instructor, his ungainly effort to match her graceful
stride made for hysterical choreography.

Falling behind the grasshoppers is ironic, given our propensity to race ahead of nature's slow dance. I spent my first summer in Wyoming decelerating, trying to pace my manic recording of grasshopper behavior to their protracted periods of rest, punctuated by contemplative walks and languid feeding. They spend nine months underground as eggs and live only a couple of months after hatching. With such a short life, you might expect a frenetic search for food and mates, but their days unfold slowly. Grasshoppers would make fine artists and poor scientists. They don't have datebooks, goals, or regrets. Their task is to be grasshoppers. The purpose of dance is to dance.

On the prairie, the band-winged grasshoppers herald the coming spring. Having spent the winter tucked under the duff as nymphs rather than as buried eggs like most species, these grasshoppers are the first to mature and take flight. The warm days of May are punctuated by the clatter of their red and yellow wings. In China, these grasshoppers perform a spellbinding dance. On a sunny afternoon, they leap straight into the air, their lemony wings crackling while they bob as if suspended on an invisible elastic thread. The dance is utterly chaotic in its timing but highly disciplined in spacing, as if each grasshopper was assigned to a one-square-yard patch. It reminds me of the only dance in which I regularly participate.

Nestled in a hillside on the snow-crusted high plains, embedded in the starlit ebony of the winter solstice, my Unitarian Fellowship heralds the coming days of growing light that will become spring. Spaced evenly by holding hands, we sing "Lord of the Dance" and weave a line that passes through and around itself. The only music is our voices; the only rule is that the line remain unbroken. The bounded chaos is a model of life and pure celebration.

To Be Honest

*To every thing there is a season, and a time to every
purpose under heaven. A time to be born, and a time to die;
a time to plant, and a time to pluck up that which planted;
A time to kill, and a time to heal; a time to break down,
and a time to build up.*
—Ecclesiastes

MY JOB IS to kill. But I usually describe my profession
euphemistically as "applied ecology" or "pest management." As an entomologist on the faculty of the University of Wyoming's College of Agriculture, I work to
develop new and better methods of managing grasshopper
outbreaks that would otherwise devastate the Western
rangelands that ranchers depend on to feed their livestock.
While agriculture brings forth life, entomology is largely
premised on taking life.

I flatter myself that I make substantial contributions
to science by refining the use of insecticides. But the bottom line is that I am an assassin: My job is to extinguish
life. I am expected to do it well—efficiently and profes-

sionally. This year I will direct the killing of no fewer than 200 million grasshoppers and more than a billion other creatures, mostly insects. Their accumulated bodies will weigh more than 250 tons and fill twenty dump trucks. That's a lot of killing, and each year it gets harder.

I began to study grasshoppers in 1986, learning how they spent their days. Few scientists have taken the time simply to observe these insects, although this seemed to me a reasonable initial step in getting to know them. During that first summer in Wyoming, I savored long, lonely hours watching grasshoppers in a field south of Laramie. As the insects lounged in the soothing morning sun or took siestas in the sizzling afternoon heat, I struggled to stay alert and focused on my work, systematically recording their behavior for later analysis. In the words of Konrad Lorenz, "It takes a very long period of watching to become really familiar with an animal . . . and without the love for the animal itself, no observer, however patient, could ever look at it long enough to make valuable observations on its behavior."

Our empathy for animals that are not soft and warm grows slowly; years of familiarity breed compassion. It might seem difficult to connect intimately with grasshoppers, but they are rather endearing when you give them a chance. We readily bond with infantile forms; the oversized head and eyes of a baby melt our hearts. It is no coincidence that Walt Disney chose a cricket for his first insect character—Jiminy's features trigger a subconscious parental instinct. Grasshoppers are innocent and affable creatures, although they can be a bit scrappy when crowding around a particularly tasty morsel (such as a recently deceased comrade).

Grasshoppers are also beautiful animals. On an afternoon's walk, I can find twenty or more species. *Dactylotum bicolor* is a garish pink, blue, and black pencil-stub of a grasshopper. Like Jack Sprat and his wife, the sleek, velvety black males of *Boopedon nubilum* are the antithesis of the obese, mottled-brown females. The size of a mouse, the wingless pink-and-green *Brachystola magna* lumbers across weedy fields munching sunflowers. Only a sharp eye can detect *Hypochlora alba*, a ghostly green grasshopper that vanishes as it tumbles into a patch of the cudweed that constitutes its only food. An adrenaline rush gives way to a sigh of relief upon finding *Aeropedellus clavatus*, an acoustical mimic of the prairie rattler. I can't help but wonder if all this splendor is a necessary consequence of evolution or whether it's simply a miraculous expression of joy.

When I started my research, I was largely an observer of the insecticidal spectacle conducted by various agencies and orchestrated by the U.S. Department of Agriculture. These programs were intended to protect the grass needed by livestock, and the USDA paid the entire cost of spraying pests on federal land, half the cost on state land, and one-third of the cost on private land. As a consequence, the killing fields were immense. The smallest operation permitted by policy was ten thousand acres, but this does not approximate the scope of effort brought to bear against a major outbreak. In 1985 and 1986, more than twenty million acres were blanketed with insecticide.

As a developing scientist new to such staggering proportions, I was eager to innovate and explore novel approaches to improving grasshopper management. To my disappointment, any interest the USDA had in new ideas at that time was subsumed by the enormous investments that fueled their control programs.

However, in response to economic and environmental concerns, the USDA was forced to abandon its subsidy program in 1996, a decision that effectively tripled the cost of grasshopper control for ranchers. This is where I came in. With my colleagues, I developed a method of application whereby insecticides are applied in strips at a fraction of traditional rates, with untreated strips in between. This wasn't rocket science, but it worked (largely because grasshoppers move readily into treated swaths, perhaps in search of a cannibalistic snack).

When we generated our first large-scale success on three square miles of rangeland, I knew something of what J. Robert Oppenheimer felt as the first atomic fireball blistered the New Mexican desert and he muttered, "I am become Death, the destroyer of worlds." Like me, he never chose a target, ordered a strike, or pulled a release lever, but he provided those who make such decisions with the option. Recently, the method that we pioneered has been employed on a quarter of a million acres throughout the West, killing nearly twenty billion grasshoppers.

Perhaps because I began relating to grasshoppers one at a time, my encounters with control programs have a haunting quality. Unable to fully grasp the number of individuals killed in a control program, the effect is similar to scanning a stadium packed with tens of thousands of individuals. It's impossible to make out faces in such a crowd; they meld into an anonymous mass of humanity, a numerical attendance figure.

To conceive of the single largest spray program in the West, imagine a desolate tract of rangeland encompassing six million football fields. This area is so large that you would need a month to walk its border and a company of gardeners equal to the population of New York City to mow its surface in a day. Over this area, we spread 375,000 gallons of insecticide, the amount of liquid that would be contained in a line of beer bottles 157 miles long—enough to give every person in the United States a heaping teaspoon. A few days later, we have killed five hundred billion grasshoppers. With this number of corpses, we could fill six hundred gymnasiums, and it would take a working surface of twenty square miles to stitch a quilt of their wings. Imagine giving each person on Earth eighty of these small corpses.

Walking across the prairie after the first large-scale program that I had witnessed, I hardly noticed the corpses, but the stillness was profound. In a grassland swarming with life one day earlier, there were now no bees playing connect-the-dots with wildflowers, no ants staggering under their masochistic loads of seeds, no ground beetles lumbering in their search of a meal, no flies circling with their whining plea for a sip of sweat, and of course, no grasshoppers erupting in pandemonium from beneath my feet. Like Lilliputian scenes from the suburbs of Hiroshima, the architecture remained, but there was no movement and no sound save the ghostly swaying of the grass and the eerie whisper of the wind. To hear the silence Rachel Carson foretold is to know the power of poison.

These days, the weapon of choice for most ranchers and local pest-management agencies is diflubenzuron, a chemical that even at very low levels inhibits the formation or hardening of the insect's cuticle, or exoskeleton. I am proud that we now use less than 1 percent of the amount of insecticide that was applied just ten years ago, but the quantitative success belies the qualitative effects. Within a week of treatment, the insects molt into deformed monsters with grotesquely twisted wings and misshapen legs. Often their hind legs fall off because the cuticle is too weak to withstand the pressure of jumping. They become lethargic, appearing dazed or exhausted. Like forgotten war refugees, the amputees stagger about aimlessly. Each summer after spraying, I walk the prairie to see the gruesome results of a control program so that I never forget what I have made possible.

The first time I fully sensed death at such a tremendous scale was during a sabbatical leave in Australia, a land known for its ability to touch the human spirit. My purpose in going there was to learn from the world's most efficient and effective locust-control program. The Australian Plague Locust Commission's headquarters bears an uncanny resemblance to a war room, with color-coded targets on wall-sized maps, the static-laden chatter of radios, and a camaraderie reminiscent of a grizzled platoon. The formal lessons they offered me concerning logistics, communications, and surveys were informative, but what I ultimately learned was not part of the structured tutorial.

The sparse grasslands that fringe the outback near Broken Hill are notorious for their capacity to foster locusts—the migratory, swarming form into which some grasshoppers develop under outbreak conditions. My introduction to these creatures occurred on an unusually wet, gray morning. My guide pointed at the birds swooping from a distant tree line to join a raucous flock just visible as it churned in the clumped canopy of grasses. It was an enormous all-you-can-eat buffet of locust nymphs.

Nymphs are the immature, wingless stage of grasshoppers or, in this case, locusts. Although each one is only the size of a pushpin, they form bands ten feet wide and a mile in length. With two thousand nymphs per square yard, a single band may include twelve million insects— roughly equivalent to the entire human population on the Australian continent. Warmed by the morning sun, the nymphal anarchy coalesces in a surging, leaderless mass, and the march begins.

On a quiet afternoon, the faint rustling of millions of mandibles grinding and bodies tumbling accompanies the sinuous band as it rolls across the grasslands. These waves of life are so thick they can be seen from airplanes; at five hundred feet they appear as arching shadows with rings of bare soil in their wake, marking the places where they stopped and fed on previous days. Deep in the outback the adults are relatively harmless, but their flights take them inexorably toward the farmlands that border the arid interior of the continent. The swarms look like shimmering dust storms rolling over the baked grasslands. Grains of sand, stars in space, locusts in flight—the sheer numerical splendor is unfathomable.

Early one morning we traveled to the site of a spray program. I had planned to lend a hand in the body count, or "efficacy assessment," but I never got beyond trying to comprehend the essence of the massacre. I had observed such programs in the United States, but never one in which death was so apparent. Perhaps it was the recency of the insecticide application, or my empathy after having been engulfed by a swarm the previous day, or my detachment from the farmers who benefited from the program, but I was stricken by the scene around me. Everywhere locusts were lying in the burnt-red dust. Some were dead, but many were still twitching in the spasms brought on by the neurotoxin. I am told that some bombardiers in World War II were unable to continue their duties once they witnessed the carnage on the ground. It was as if I had really seen, for the first time, what it meant to dole out wholesale death.

When I was in fourth grade, my older brother, who occasionally thrashed me when we were younger, came to my defense against a seventh-grade bully. The miscreant never again accosted me, but my brother did. As the love of a brother transcends the scraps and tussles of growing up, so my relationship with grasshoppers is conflicted, swinging between affection and aggression. In the midst of my own violence, I have defended them from government bullies who understand little of their nature and care even less.

In 1989 the USDA planned to import exotic parasites and pathogens to attack the native grasshoppers of the Western states. These organisms, tiny wasps that lay their eggs in those of grasshoppers and fungi that invade the tissues of insects, were to be gathered from around the world and released on the prairie. Because it risked permanently damaging our vast grassland ecosystem, I engaged in a long, bitter, and eventually successful battle to stop this effort.

The program was missing important elements, such as a provision for protecting nontarget organisms. I found results of laboratory studies hidden away by the USDA revealing that the wasp proposed for release would have parasitized a grasshopper species that suppresses snakeweed, a poisonous plant that induces abortions in cattle. Yet there would have been no way to put the genie back in the bottle if things had gone dreadfully wrong.

At least with insecticides, every use is an intentional act that invokes regulatory constraints. We can avoid national parks, wildlife refuges, and habitats where endangered species reside—unlike parasites, which spread and invade without regard to such considerations. And we know that chemicals will eventually break down, whereas parasites and pathogens will increase in numbers. For all of the damage that we can do and have done with toxins, the potential for ill-conceived introductions of exotic species to disrupt natural systems is even greater.

Grasshoppers, unnoticed, sculpt the prairie, prune poisonous weeds, compost dead plants, and feed the birds. And with each passing year, the quiet chorus of their lives becomes more deeply instilled in my being, while the rousing refrains of their sporadic outbreaks continue to draw the attention of government agencies.

Taking life, like giving life, can be a sacred act. A friend of mine who is an herbalist believes that one must harvest medicinal plants with thankfulness, understanding, and humility to access the full potential of the resulting extract. Indigenous peoples asked permission or forgiveness of the animals they hunted, and perhaps modern agriculture should act with such humility and grace when killing is necessary. Following the lead of native people, agriculturalists would do well to understand that the land is shared with other creatures and that their needs are worthy of our understanding.

At the beginning and end of each summer, I sneak away from my field assistants and collaborators to be alone, to pray. This is a time when I experience the fullness of the prairie, when I seek what lies at the core of my intentions as a scientist, and when I release the guilt and shame. The thought-words are different each time, but the question I ask myself persists: Why do I continue to develop the means of killing these creatures?

I justify killing grasshoppers because my intentions are purified by love for them. I am soothed by the notion that I mean well, that I foster a world in which there is less killing and fewer misunderstandings between species. I tell myself that intentions are all that we really control; outcomes are evasive and uncertain. But spraying thousands of acres with insecticides, regardless of intentions, is going to do a lot of harm. A Buddhist priest once told me that the Samurai were Zen masters, that they killed with a depth of awareness we can barely imagine. The mindfulness of Buddhism allows one to be profoundly effective, but, he noted, you are still stuck with deciding what to do effectively. Slicing an enemy in half with perfect awareness still makes a mess.

Vignettes from a documentary on killing have haunted my memory for years. The film followed a condemned killer over the course of his last days. Ironically, I found the most compelling character to be not the convict, nor his family, nor the victim's family, but the warden. Here was a good and kind man burdened with the obligation of premeditated murder. The warden acted with dignity and compassion; he was gentle but not fawning, supportive but not patronizing, regretful but not apologetic. He struggled to make the most difficult of all social responsibilities as decent as humanly possible.

And so I ask, Who is taking the more meaningful role: the compassionate warden at the side of the condemned, making the execution more dignified and decent, or the protestors in the streets outside, shouting slogans against the death penalty? Perhaps I would be a better environmentalist if I refused to use insecticides, but would I be creating a more decent world for my fellow creatures if I left my job in protest? Or should I keep working to find ways of killing fewer creatures with more humane, less toxic methods?

I claim to be a compassionate executioner. After all, if somebody has to kill, it might as well be someone with an inkling of empathy. But to be precise, I am not the executioner; that would be the aerial applicator. In fact, I generally avoid the treatment programs. To sustain the analogy, I'm really in the business of designing less cruel electric chairs, and this sometimes seems even more perverse than pulling the switch. Perhaps it doesn't do the convulsing grasshopper (or inmate) a whole lot of good to know that I care, but maybe society can retain some of its humanity if I can make the act of killing less brutal and less frequent.

Inescapably, we live by death. I study how to kill better and less, but a little more of me dies with each field season. If my current projects succeed in reducing the amount of toxin needed to restore grasshopper infestations to naturally manageable levels, during the next outbreak we will reduce our insecticide use by ten thousand tons and limit our killing fields to less than half of their historical scale. Still, the carnage will not be avoided; that is the ugly middle ground called compromise. I am convinced that it is easier to be a member of Earth First! or the Chemical Manufacturer's Association than a member of neither.

I sometimes wish I could throw myself at one of the extremes: environmentalism or anthropocentrism, mysticism or rationalism, religion or science. But to do so is to become truncated, half human. Some people can choose one or the other, but living and working with grasshoppers, knowing their beauty and innocence while being deeply responsible for their deaths, has shown me that both of these ways of knowing must be honored for our agriculture and our civilization to flourish.

Scientists have a unique opportunity in this venture: We can understand, connect with, and tell the story of the natural world. For those who are willing, it is possible to both comprehend the facts and transcend them for the sake of the spirit. And when we are open to this possibility, I have found that with time, technical data give way to a deeper kind of knowledge that relies on intuition, tradition, experience, and faith—and beyond knowledge lies the possibility of wisdom. For me, the data of "percent mortality" has given rise to the knowledge of how to control grasshoppers with minimal harm. My wisdom, however nascent, comes from seeing the death of grasshoppers and the integrity of the biotic communities (including the human elements) and realizing that we are all one, that we are diminished by their deaths and uplifted by their lives.

Scientists often attack spirituality, and I believe that the reason is fear—a sense of anxiety, even dread, that there remain elements and processes of the world that are fundamentally and ultimately beyond our capacity to control, quantify, or rationalize. I suspect that many scientists are hostile to mystical notions in order to prevent the impure thoughts of transcendent experience, the troubling touch of subject feeling, and the unsettling whisper of spiritual insight from confusing the rational process of analysis. To focus on one-half of what it means to be human, they chant a mantra of materialism to deny anything that lies outside of science. Rather than sustaining the illusion of objectivity, they could open themselves to a respectful, caring, even loving relationship with the creatures they study. But then they would end up like me—attached to the creatures I kill, with all of the unrest that this entails.

Although ideals make for fine philosophical tracts and political treatises, real life is full of complexity, uncertainty, diplomacy, and nuance. I often encounter those who preach about how the world ought to be, but rarely do these futurists wish to engage in the actual work of getting us from here to there. Of course, the middle way risks the pitfall of rationalization, as we talk ourselves into perpetuating the status quo. I try to forgive myself and secure nature's pardon by contending that with each killing season, we are one step closer to a just and compassionate future. And so I will continue to decimate grasshoppers for the benefit of agriculture and continue to mourn them for the sake of human culture.

But putative virtues of "killing them softly" beg the question of why kill them in the first place. There are no comprehensive solutions to the universal dilemma that we must kill to live. No philosopher, theologian, artist, or scientist has offered a general solution to the mystery of suffering and death. Paradoxes are not solved; they are lived in and perhaps lived through. The wicked irony embedded in this paradox is that I take psychic sanctuary in making my victims anonymous by numerical imaginings. On a recent trip to Australia, I watched an aerial spray program with detached interest, taking mental notes of logistical and methodological details. But I cringed when a colleague looked down at the parched soil and lifted his foot to grind a single nymph under his heel. It is harder to crush a grasshopper slated to be sprayed than it is to watch the airplane lay down a blanket of death.

My memory is drawn to the summer of 1992, when I stood on a desolate tract of sunburned rangeland infested with Mormon crickets (which are actually a bizarre species of katydid named for their legendary invasion of Mormon farms) near Edgerton, Wyoming. A dense band of insects numbering in the tens of thousands had been sprayed a few hours earlier, and I arrived to find them staggering about on the slopes. Humpbacked, wingless, with antennae streaming behind, it was as if a sadistic fisherman had dumped a netful of melanic prawns on the prairie, where they flailed desperately.

The poisoned crickets were unable to hop uphill, so each movement brought them down into the ravines. They accumulated in the dry creek beds in what looked like stagnant black streams. On the hillside I lifted one cricket from within the sagebrush where it had become entangled. I grimaced at its convulsions, which were aggravated by my handling, then set it down on the dusty soil to resume its dying journey to the anonymous mass of writhing bodies at the bottom of the draw.

So in the end, how do I live as an assassin? I know that the grasshoppers' suffering and my pain are real. I know that they die so I might live. Grasshoppers are my ecological communion; their bodies are my life. Through them I have found meaning in my work, experienced connectedness to other beings, and gained a sense of purpose in this world. Perhaps my destiny is that of the warden, to ensure that these creatures do not die unknown by the hand of a dispassionate executioner. To be mourned is to have one's life—and death—touched by another sentient being. Perhaps that is all that any of us can hope for.

But does one man's perspective offset the billions of deaths? I am suffocating under the expanding mass of corpses that pile onto my conscience each year. And so, I tell their story and mine—and ask something of you. At your next meal, say grace, give thanks, remember them.

From the Mouths of Babes

*We can't form our children on our own concepts; we must take
them and love them as God gives them to us.*
—Johann von Goethe

I DON'T LIKE TO think of myself as an executioner.
I prefer to be called an economic entomologist. My job
is to kill insects and other creatures. Applying pesticides
may be the only activity less socially acceptable than
chasing ambulances. So I thought that showing my kids
an authentic rangeland-grasshopper outbreak would
help them make sense of my work.

Ethan struggled his way up into the cracked vinyl
seat of the Chevy pickup, his cheeks flushed and eyes
wide in anticipation. Kindergarten had been fine, but
summer was better. Some 125 miles and 125 questions
later, we arrived in Guernsey, Wyoming, and stopped
at Katie's Diner for lunch. My job was already impres-

47

sive to him—I got to drive a truck and eat lunch in a diner.

The screen door of Katie's banged shut behind us. We climbed into the truck and headed to Whalen Canyon, a two-thousand-acre tract of severely infested rangeland a few miles north of town. Although early in July, it was nearly 100 degrees, and the wind conspired with the heat to create the equivalent of a convection oven. After a short walk in which we flushed waves of grasshoppers from the boot-high bromegrass, I tried to explain my job.

Squatting on my haunches, a posture Ethan instantly mimicked, I began, "You see, pal, all of these grasshoppers are eating the grass. And the rancher, Mr. Martin, needs the grass for his cows."

Ethan stared into the distance, either spellbound by my lecture or imaging the frosty Coke that I'd promised him when we got back to town.

"We're trying to find a way to get rid of some of these grasshoppers. We're testing a new chemical, a poison really, but it won't hurt the other animals as much as what we use now."

Ethan poked at a grasshopper that was woozy from the heat and desperately clinging to a stem of grass in hopes of avoiding the searing soil. I went on, trying to show him that if I wasn't exactly a hero, neither was I a villain. "The grasshoppers are part of the prairie and we don't want to hurt them, but we're trying to live here too."

I finished with a story about the pioneers that passed just five miles east of Whalen Canyon on the Oregon Trail, explaining how they had seeded this region with the ranching families and how the locust plagues nearly drove them from the land. Mr. Rogers could not have given a better accounting of economic entomology. Confident that Ethan saw the honorable character of my work, I stood up, a bit too fast. A wave of vertigo washed over me as Ethan offered his analysis.

"Dad?" he began, oblivious to my condition.

"Yes," I replied.

"Weren't they here first?"

On the drive back to town, my rambling discourse touched on indigenous people, native species, property rights, and why calling "dibs" did not establish a moral position, but I have yet to really answer Ethan's question. At least I was forewarned when I took his sister to the field a couple of weeks later. That was good because Erin was two years older and intellectually precocious.

This time I'd introduce the principle of sharing to pre-empt the ethical questions.

By now, the rangeland grasses were baked to the color and consistency of shredded wheat. As we picked the tiny spears that constitute bromegrass seeds out of our socks, I delivered a brilliant lecture on Western agriculture, being sure to discuss ways of fairly distributing resources between humans and nature to ensure coexistence. Erin asked me how many different kinds of grasshoppers there were in Wyoming (112 species) and how insecticides kill (malathion and related compounds short-circuit the nervous system). She seemed satisfied, even intrigued; I was delighted. After our Cokes in town, we headed back to Laramie.

"So," I said in an effort to summarize my work, "you see that there just isn't enough grass to feed both the grasshoppers and the cows. Sharing is ideal, and we try to make it work whenever possible, but sometimes there just isn't enough to go around. And this year we have to kill some of the grasshoppers to be sure that the cattle have enough to eat." I looked down the road, watching the heat rising like an undulating, liquid curtain over the asphalt.

"I see," she offered, with that quaver in her voice that I have since come to recognize as being the equivalent of the deepening whistle associated with falling bombs in war movies. "And next year," she continued, "will it be the cows turn to get killed?"

We Murder to Dissect

Sweet is the lore which Nature brings;
Our meddling intellect
Misshapes the beauteous forms of things—
We murder to dissect.
—William Wordsworth

"I CAN'T FIND the dang heart in all this mess," the student pleaded while poking ineffectually at the mangled body.

After thirty years of dissections, the sight of an animal splayed open still makes me cringe, at least inwardly. "Don't try to cut your way through the viscera, just push it aside to get to the dorsum," I advised. The insect's heart is a thin vessel that runs down its back (where a spinal cord would be if insects were vertebrates) and extends the length of its body—an unfamiliar, even alien, morphology when encountered for the first time. Our first reaction to insect bodies is typically dissociation. They lack lungs (insects breathe via a net-

53

work of diaphanous tubes delivering oxygen directly to the tissues), a urinary tract (insects have tiny tubules that pass solid uric acid crystals into their alimentary canal to be mixed with their feces), and a circulatory system (insects have a heart, as the student eventually discovered, but they lack arteries and veins; the blood simply sloshes around, with the heart functioning like the pump in swimming pool). Learning the structure and function of the organ systems is the proximate goal of our course Insect Anatomy and Physiology, but for the student to truly see and understand these awful/wonderful creatures, they must ultimately overcome their sense of alienation.

Even with a familiar animal, students easily forget that they are systematically tearing apart a once-living organism. During my high-school years, Albuquerque established the Career Enrichment Center, a citywide facility for advanced studies in a range of disciplines, including biology. In this setting, I took a class called The Human Animal, which cemented my devotion to the study of biology and illuminated the power of dissection as a learning method. Each student was given a cat—furry, soft, unique (mine was a tabby), and dead. The initial sense of connection was quickly lost as the animals were flayed and meticulously examined over the course of weeks. Struggling to find and memorize muscle origins and insertions in a tangled maze of tissue converted the creature into an object, a pungent puzzle. In the terms of Martin Buber, the work of science transformed the I-Thou relationship between beings into the I-It interface between subject and object.

But with the dissection of the stomach, a powerful reconnection developed. Each student peered into the last meal of a cat, its final act of feeding, the undeniable evidence of its having lived. My cat was chock-full of masticated, dry cat food, but my bench-mate's cat had the remains of a mouse in its stomach, including a furry little leg that had been swallowed whole. With these discoveries came a glimpse of the animal's life, a clue to its story. The last supper of the condemned grips our souls and imaginations. The final meal of a death-row inmate reminds us that whatever distance has been created by a horrific deed—a distance so great that we can bring ourselves to kill in retribution—we ultimately share the common experience of savoring a meal.

Not all dissections achieve the ideal of connectedness. In fact, most exercises fail miserably in terms of even their proximate purpose, learning the anatomy of other life forms. In my "regular" high-school biology course we dissected a frog. Armed with dull scissors and useless plastic spatulas, we searched for various organs with sporadic success through the remains of a pickled leopard frog. The entire process lacked the essential elements of intimacy. It was hurried; we had forty-five minutes from initial incision to clean-up. It was public; we worked in haphazard trios, hastily composed following an order to "get into groups." And it was impersonal; growing up in Albuquerque, I'd caught lots of lizards but had never even seen a live frog in the wild. This is not to say that dissecting frogs, the classical drill in biology classes, is necessarily a meaningless exercise. In fact, this was my second experience of this sort, and it could not have been more different from the first.

My first dissection of a frog was magical, transforming my understanding of living creatures from one of mere curiosity to that of absolute wonder. When I was in the fifth grade, my parents gave me a biology kit for Christmas. Inside the box was a detailed workbook, a shiny set of dissection instruments, and a series of jars containing animals of increasing complexity, each in a precisely cut-out compartment in the Styrofoam packing. The manual began with the dissection of a starfish, followed by an earthworm, advancing to a grasshopper, a perch, and finally a frog. The invertebrates were not well preserved, and the tools were not particularly effective in working with such small creatures. The perch was a disappointment, being jam-packed with eggs that displaced and obscured most of the organs. But with each dissection, I learned a bit more about both the fine-motor skills of dissection and the similarities between other beings and myself. I studied the diagrams of the frog for days, thrilled with the possibility of actually touching a real liver, an actual lung, and a once-beating heart. The drawings made the frog look like a tiny person, with all of the same organs in the same places that ours would be found. I set up the dissection on a card table in my room after Saturday dinner. Closing the door, I began by making a careful cut along the length of the frog's belly and gently laid back the thin skin to reveal the marvels within. Transfixed by the colors, textures, and exquisite details (I managed to find the chambers of the heart), I spent the evening in rapt wonder. My own body had these same organs. The paradoxical sense of life's fragility and vigor has never left me. The spell was finally broken by my mother's muffled declaration from the other side of the door of an impending bedtime. If my first dissection was the definitive experience for a ten-year-old boy, my first vivisection was transformative for a twenty-year-old man.

Being a biology major at New Mexico Tech meant taking Animal Physiology, a course tailored to the needs of my premed classmates. Taught by Dr. Smoake, the premed advisor with a remarkable record of placing students in medical schools, we dreaded the course. Dr. Smoake was a demanding, uncompromising, and tough professor of the old school, and you either learned a whole lot or changed majors. The lectures were brilliantly delivered but wickedly paced, the exams were fair but fiendishly difficult, and the labs had a scheduled starting time (which nobody dared miss) and a putative ending time (which nobody dared notice). Most of the labs involved experiments with anesthetized rats, and there was one particularly successful study of the cardiac function of turtles. But the most complex and exhausting lab involved working with feral cats that were destined for euthanasia at the pound. They were so wild that Dr. Smoake and his assistant delivered the injections that would render them unconscious through the burlap bags that held the cats. The anesthetic allowed us to work for a couple of hours before a lethal dose was administered.

We all understood that the cats would never feel the pain of our cutting or suffer the consequences of our inevitable mistakes during the vivisection. We had worked on rats, but the cats were somehow different: They had not been bred for our use. They could have come to know play and affection, and with time they might have trusted people to care for them. Our usual banter was replaced by hushed, respectful tones. There was a sense that we had to extract every possible bit of knowledge from this exercise to justify the use of these living creatures.

The irony is that I can't even remember what physiological system we studied that afternoon. I know that we used cats because we were working on some element of the body that was impractical to study at the scale of a rat. I vaguely recall teasing a ureter free from the surrounding connective tissue, but that is all I retained of the proximate lesson. What floods my memory was the near-holy atmosphere of profound respect that Dr. Smoake promulgated by the unspoken example of his own intensity and manner that day. Today such a laboratory experience would be impossible at most institutions, but twenty-five years ago different rules prevailed. And I am not sure that today's biology majors are better off intellectually, emotionally, or spiritually for their carefully sanitized and controlled experiences. Nor am I convinced that our fellow creatures are necessarily better off for our efforts to educate students at a distance from death—and ultimately from life.

If dissections can deepen our sense of connection and responsibility, they have corresponding power to distance students from the living world. And such alienation is facilitated by using those creatures least like us. In my graduate training and, to my shame, among my own colleagues, I have silently witnessed and thereby countenanced the vivisection of insects without the relief of anesthesia. The dissection of a living cockroach to demonstrate the beating of its heart (and the various effects of temperature and selected chemicals on cardiac function) is common in entomology laboratories. In cases I have seen, the insects were immobilized by a few minutes of cold, then pinned to the dissection tray, their legs removed (otherwise they flail about when the insect warms and its capacity for movement is restored), and the abdomen is eviscerated to expose the pulsing heart. The stumps of the legs wiggle frantically as the instructor either conducts or oversees the callous spectacle. The implicit and powerful message concerning the creature is that its suffering is either nonexistent or not worth our concern. The insects are already foreign to us, having antennae and an abundance of legs while lacking warmth and fur, so the act of cutting open these animals without the benefit of anesthesia simply adds stones to the wall of separation between us.

There is evidence that insects feel pain, and the intuition of most students is that all creatures are somehow sentient. We cannot, of course, be certain, but in the face of even a remote possibility that these animals suffer, surely we are obligated to mitigate their potential pain. And so I have come to ask my colleagues to either anesthetize the creatures or quickly decapitate them, as either method allows the heart to continue beating while providing relief from suffering. However, even if these simple procedures are adopted, they provide only the veneer of compassion. These methods do not ensure that the students are drawn into a sincere sense of obligation to, or authentic understanding of, the animals lying beneath their forceps.

The faculty who team-teach Insect Anatomy and Physiology with me convey a sense of childlike wonder during the dissections. While I struggle with my sense of moral ambiguity, their murmurs of encouragement to the students and their candid exclamations of delight reveal their genuine and undiminished (one of my colleagues is nearing retirement) amazement with the creatures that they have devoted their lives to understanding. For me, the connection to the mangled grasshoppers is more personal. We could purchase them from a scientific supplier, but each summer I collect immense, mouse-sized grasshoppers from the weedy roadsides of Wyoming. This summer a stretch of sunflowers along a dusty road harbored hundreds of these pink, tan, and green flightless behemoths. With the help of my daughter, I gathered dozens of these clumsy and rather repulsive insects: They immediately defecated and regurgitated upon being caught, while kicking menacingly with their spiny hind legs. They were put into a cooler of ice and then frozen in my lab for future use. In pragmatic terms, our frozen specimens are far superior to the chemically preserved bodies provided by the supply houses, as ours retain the textures and fine structures of the various tissues. The students feel the greasy slickness of the fat body, the stringy tenacity of connective tissue, the rubbery toughness of the brain, and the diaphanous sheerness of the tracheoles (breathing tubes). But these matters alone do not motivate my annual collection.

When I have gathered the grasshoppers for the students, the insects are real, their source is known, and I become responsible for ensuring their meaningful use. I can tell the story of their lives and how they came to the hands of the students; I don't know the story of the Eastern lubber grasshoppers that come in jars of preservative. In the frigid days of February, I can tell the students how the brown stains of the grasshoppers' regurgitated meals remain on my hands for days, how the insects bask on the leaves of roadside sunflowers to gather the heat that drives their development, and how such conspicuous insects seem to avoid predation by being distasteful.

But whatever initial sense of connection that may develop is quickly lost as grasshoppers are flayed and meticulously examined. Struggling to find and memorize muscle origins and insertions in a tangled maze of tissue converts the creatures into mere objects. With the dissection of the stomach (technically, the foregut or crop), the students peer into the last meal of the grasshopper, the undeniable evidence that it lived. Often the gut contents are macerated plants, with a few containing the shredded wisps of grasses or the remnants of seeds, rather than a bolus of sunflower leaves. Sometimes a student finds an empty gut, which suggests a newly molted individual who hadn't had time to eat while its new body hardened. And every year at least one student finds the remains of another insect, often the leg or foot of another grasshopper that has been swallowed whole. Grasshoppers, especially females in search of lipids for their developing eggs, supplement their diet with the scavenged remains of their brethren. With these discoveries comes a glimpse of the animal's life, a clue to each one's story. The anatomy of these animals may seem genetically standardized, but their last suppers reveal that the flayed bodies lying pinned to wax-bottomed trays are those of unique individuals.

In my Great Books of the Life Sciences course this semester, we will encounter H. G. Wells's Dr. Moreau —the ultimate mad scientist. When Edward Prendick becomes marooned on the island of Dr. Moreau, he soon realizes that the creatures he encounters are "not men, had never been men. They were animals— humanized animals—triumphs of vivisection." Of course, the horribly mutilated creatures are not fully humanized and eventually turn on their creator, who admits to Prendick, "You cannot imagine the strange colorless delight of these intellectual desires. The thing before you is no longer an animal, a fellow-creature, but a problem. Sympathetic pain—all I know of it I remember as a thing I used to suffer from years ago." The wicked irony of Wells's story is that the justification of dissection or vivisection lies not in making animals more human but in making *ourselves* more human—and more humane. The abysmal failure of Dr. Moreau was that he intentionally objectified the creatures in order to satisfy his own perverse scientific agenda.

To cut into an animal, whether on a laboratory bench or at the dinner table, is to partake in its death, to accept that it was killed for you, to obligate you to securing physical, intellectual and spiritual sustenance from its life. If done well, the act of dissection creates the moral demand that students learn. At the least they will learn anatomy, but in twenty years they are unlikely to recall the position of the subesophageal ganglion. Perhaps if I do well, they will remember the elegant complexity of insects, recall the individuality of grasshoppers, and know that all life—even that which seems hard, cold, and alien—is worthy of our respect.

Tough Love

*It is far from easy to determine whether she [Nature] has proved
to man a kind parent or a merciless stepmother.*
—Pliny the Elder

"THAT'S ALL YOU'RE WEARING?" I asked, trying
to hide my astonishment. Cutoffs were acceptable, bra-
less was fine, but this was a real problem.

"Yes," Carlie replied with a note of defiance. Her
spunk had seemed like an asset when I'd hired her as a
summer field assistant.

"You don't have any shoes or boots in your pack?" I
tried hopefully, pushing up the front of my hat to scratch
my receding hairline. After a two-hour, early-morning
drive from the university to the field site, I did not relish
the thought of canceling the training session.

"No. These are fine. I've hiked miles in them," she
gestured at the leather sandals. "OK," I relented, "but

this isn't a mountain meadow. Be careful and go slowly."
We managed to get in a couple of hours of practice in
the methods of sampling rangeland grasshoppers before
a prickly pear cactus terminated our session. Within the
next month, I would need reliable workers to track the
results of our experiments in pest management—and
Carlie would need a pair of leather boots.

For Carlie and most people, the prairie connotes rich
soil, wildflowers, and tall grasses, not the harsh, spiny
grasslands of Wyoming. On the rangeland north of
Cheyenne, the grass is typically knee-high. Needle-and-
thread grass, named for its seed, characterizes the hostile
flora. The seed is a half-inch spear trailing a three-inch
spiraled filament, an exquisite dispersal adaptation that
turns grazers into mobile pincushions. At least humans
have the dexterity to extract these tiny harpoons from
clothing before they penetrate tissue; livestock can develop horrible abscesses around the embedded seeds.
Livestock can also fall prey to the lure of snakeweed, a
plant that stays temptingly verdant—and poisonous—
even when the rest of the prairie is baked to a crisp.

Introduced plants have joined in the campaign to keep
Wyoming's prairies inhospitable. Cheatgrass, brought to
America by European settlers, earns its name by pro-

viding a lush blanket of green for a few days in early spring, which rapidly matures to a purplish brown, like a week-old bruise on the prairie. For the rest of the year, these unpalatable stands frustrate livestock and wildlife and torment my field assistants. The seeds are tufted darts that convert socks and bootlaces into prickling, furry masses. The stands often host a fungus, whose millions of black spores cover clothing and equipment in a living soot. We want nature to be as comforting as soft cotton; the prairie offers us steel wool.

The Wyoming prairies are violent places, indifferent to our sensibilities. Carlie had missed the sight of bloody lambs strewn like rag dolls, but she was repulsed by a rancher's angry retribution—the bodies of three coyotes twisted into a fence as their viscera rotted and their limbs dried like jerky. Perhaps that image was still with her when we came across another macabre scene later that summer. "Who would have done that?" Carlie asked, her face screwed into a mix of disgust and fear. While she furtively searched the prairie for the psychopathic culprit, Scott fingered the tiny corpse impaled on the fence. Having spent a decade working with me as a research associate and most of his life in the outdoors, he knew brutality was not the exclusive purview of humans. "A shrike. He's got a whole string of 'em," he added, gesturing down the strand of wire, where every fourth or fifth barb held a grasshopper. This bird impales its prey for safe storage, and barbed wire was a fine alternative to the standard thornbush. Walking along the fence, we became desensitized to the gruesome sight, but even Scott winced when we came across the lizard hanging limply from a barb.

The amoral savagery of predators echoes the lethality of the prime killer of the high plains—the weather. Although fierce winter storms are infamous, I have been chilled by a sudden, spring snowstorm in the first week of June, peppered by fusillades of gravel propelled by shrieking winds the next week, and roasted by triple-digit heat in the last week. But as Scott nearly discovered a few years ago, the truly deadly summer weather arrives in July.

Drinking a cup of coffee at the diner in Torrington, I watched the thunderheads pile up over the rolling hills to the south of town, where Scott had been checking our grasshopper control plots. Lightening is a rare but fickle assassin, so I was relieved to see the university pickup pull into the parking lot. Sitting down on the orange vinyl seat across from me, he poured a cup from the pot on the table. "I haven't seen hail like that before," he said, looking a bit stunned. "Bad, huh?" I replied. Scott was not easily intimidated by the weather. "The truck isn't damaged, but it was comin' down the size of marbles and getting bigger as I headed into town."

When we went out the next day, the rangeland looked like a battlefield, complete with shredded shrubs, flattened grasses, and craters that attested to the barrage of icy cannonballs. Nearly a third of the grasshopper population was crushed in the storm, and at the edge of our site, the rancher found a pronghorn antelope pounded into the prairie, beaten to death by the baseball-sized hailstones. Maybe 1999 was the year for pummeling.

That year, Scott's younger brother, Spencer, joined the field crew and beat an animal to death. Given Spencer's calm, even taciturn nature, his startled yelp was distinctly out of character. By the time I turned to see the cause of his alarm, he was stomping viciously into a sagebrush. Following a final, brutal stomp, Spencer's odd assault on the bush became explicable. He reached into the gray-green shrub, grabbed the snake near the tail, and yanked viciously, tearing the body from the head that was pinned under his foot. Our rule is that we leave rattlesnakes alone unless they are in our study plots, where we are likely to encounter them repeatedly. The usual method of dispatching them is to employ a shovel, but this one had struck at Spencer without warning. He had taken it personally.

It's difficult not to take the violence of the prairie personally. The plants, animals, and weather are so harsh that we infer malevolence. But the most brutal assault is deeply personal. The scale of these grasslands assails our sense of self. You can turn away from the edge of a canyon or step down from a mountaintop, but on the prairie, everywhere you turn, the world stretches to the horizon. Being adrift in a sea of grass that is absolutely indifferent to their existence drove some early pioneers insane. In response to this assault on our ego we seek control, counterattacking through the weapons of human technology and economy.

Ranching is not a gentle art. The tools include barbed wire to control livestock, branding to control thieves, bullets to control coyotes, and poisons to control grasshoppers. My job is to develop more sustainable pest-management practices, but they are often just as brutal. A grasshopper dying as a pathogen ravages its tissues is probably no better off than one poisoned with a neurotoxin. If agriculture is a battle, mining—Wyoming's most profitable industry—is open warfare. Explosives and gargantuan machines rip open the landscape. Coal mining is a harvest of death, in which we tear open the tombs of ancient plants and animals in order to cremate their remains in our power plants. Tourism, Wyoming's other major industry, sweeps visitors past the high plains on interstate highways, disgorging them in the forests and meadows of our cool mountains. To a society that loathes suffering, a day on the grasslands of Wyoming is unmarketable.

Our cities have become dangerously crowded and our jobs coldly efficient, so we seek sanctuary in nature. Desperate to be nurtured, we attribute motive and intentionality to natural phenomena. We need to be soothed, embraced in a gentle glen. Hard landscapes that we cannot subdue, we avoid, softening them in our imaginations. So when the prairie seems to defy our idyllic fantasy and spurn our affections, we impute menace and malice. I am told that a child seeks love but prefers punishment to indifference. In a perverse way, a beating acknowledges our existence and connects us to another. But how long can we sustain this interpretation of Western lands? What about Carlie and those of us who live in this place? Must we either strike back or take our beating?

I often need to stop on the prairie to rest, or write, or observe, thereby creating the problem of posture. Standing is tiring. Squatting would work, but rising from this posture in the summer heat invariably induces a vicious bout of dizziness. Sitting or lying are made impossible by the ragged rocks, searing soil, and piercing plants. And so I kneel. In this way, I need only find two small pockets of accommodating earth for my knees. A tender mother would have us snuggled in her arms; a vindictive father would have us prostrate at his feet. Kneeling implies deference but not servility, humility but not defeat, respect but not fear. We kneel to pray, to communicate with that which transcends our personal being, to acknowledge that which is powerful and worthy, to engage the full breadth and depth of the land. And so kneeling is a good way to be on the Wyoming prairie, neither cowering like an abused child nor strutting like a conquering general.

Like Father, Like Son

Because no battle is ever won he said. They are not even fought.
The field only reveals to man his own folly and despair, and
victory is an illusion of philosophers and fools.
—William Faulkner

LIKE MANY FATHERS, my dad traveled as part of his
work. Unlike most fathers, mine was not allowed to tell
us exactly where he was going or when he'd be back.
We all knew, more or less, what a physicist was doing in
the middle of the Nevada desert in the late 1960s. It's
hard to hide a nuclear explosion, even beneath one thousand feet of earth. As I consider my own work, it seems
that conducting secret tests on weapons of mass destruction in desolate places has become something of a
family tradition.

"Don't touch it!" I warned. My daughter pulled back
her hand. "You'll get a nasty shock," I explained. Erin
surveyed the strand of electrified wire enclosing a patch

75

of Wyoming's austere rangeland. "Why did you put up the fence?" she asked. It did seem an absurd safeguard on a landscape devoid of any sign of human life other than a distant ranch house and the dusty tracks that led through the cactus, snakeweed, and sparse prairie grasses. "To keep the cattle out of our test area," I explained as we walked along, flushing waves of grasshoppers in front of us. "But why can't they go in there? There's not much grass out here for them," she replied. Erin was right; there was more grass in the enclosed pasture, but it was forage that could never enter the human food chain. "Because the chemical we sprayed in there to kill the grasshoppers hasn't been approved yet, and we can't risk having Mr. Oliver's cattle eat the grass," I offered. Explaining my work as an applied entomologist was a lot more challenging than I had initially imagined. As she bent down to watch a pliant, pale green grasshopper struggling to extricate himself from its crisp, outgrown cuticle, she asked, "What chemical is it?" "Well," I replied, "I can't exactly say." "How come?" she asked poking at the grasshopper. "Because it's a secret," I answered. At least she didn't ask to explain why companies require professors to sign secrecy agreements for this sort of work.

High-energy particle physics had a primal appeal to me as an eight-year old. Although smashing things together sounded pretty neat, the real excitement wasn't so much what my father did as the secrecy that surrounded his work. My mom was amused by the security precautions that precluded my father's divulging when he'd return. She would just read the *Albuquerque Journal* and figure it out for herself, piecing together her experience and the paper's stories about the impending and then actual detonation of another atomic bomb at the Nevada test site. But for me, the term *national security* was pure excitement. The euphemisms my dad used dripped with intrigue. He referred to *test shots*, not nuclear explosions and to *the device*, not the bomb. The best part was the code names. I vividly recall the test shot code-named *Ming Vase*; it conveyed a delicious sense of exotic mystery in the era of James Bond movies.

At first glance, the U.S. Department of Energy's Sandia National Laboratories would not seem to have a great deal in common with the University of Wyoming's College of Agriculture. But perhaps there are more similarities between my father's employer and my institution, nuclear bombs and pesticides, nation states and multinational corporations, and his enemy and mine than I could have imagined when I was hired as an assistant professor fifteen years ago. Certainly we both struggle with how to tell our stories to our children and ourselves. It's tempting to turn them into a screenplay for a James Bond movie. The unambiguously bad guys are blown to bits, but the gory results are not graphically portrayed. This sanitized version of reality creates the illusion that we can drop bombs and spray poisons without immense suffering. But in the end, our children will know otherwise.

I was eleven years old when I saw and touched Fat Man. The smooth, rotund form was slightly comical, almost jolly in its obesity. Such is my politically incorrect memory of encountering the first atomic bomb dropped on Japan. It seemed to have been named to belie its purpose, as if by hurling a schoolyard insult we could mock its power and dissipate our guilt in having built it. Of course, I saw only a model of the bomb, but I found it impossible to imagine the force, the power, the fire contained in a shell not much taller or longer than I. The second bomb, Little Boy, looked more like a torpedo, lacking the soft roundness of its predecessor. In my memory, I nearly could have wrapped my arms around its elongated form. I could have hugged a weapon that killed and maimed 150,000 people. The display at the Atomic Museum on Kirkland Air Force Base (one of the only places on the base that allowed public access) also showed how the bombs grew progressively smaller and more powerful. There was a warhead the size of my kitchen trashcan that could decimate a city; the entire class of bombs had been given some prosaic and forgettable name. Apparently, once we were able to mass-produce nuclear bombs, we quit seeing them as individuals.

Today we have insecticides named Force, Counter, Attain, Preclude, Warrior, and Fortress. The names evoke a sense of unapologetic dominance—a shameless power that feeds a constantly increasing efficacy. In the era of DDT, it took perhaps ten pounds of chemical per acre to control a grasshopper infestation; in the 1980s it took about half a pound of malathion, and now it requires an ounce of diflubenzuron. A five-gallon jug of the latter insecticide can kill more than 90 percent of the grasshoppers on an area of three square miles, a population of 100 million individuals. An insecticide that I tested in Wyoming is now being used for locust control in Africa, Asia, and Australia at minuscule rates. It's called Adonis. If we call intercontinental ballistic missiles "peacekeepers," is it any wonder that we name poisons after gods of beauty? I can now hold in one hand enough insecticide to kill more locusts or grasshoppers than there are people on Earth.

As a kid, I could not reconcile the size of atomic bombs with the magnitude of their devastation. I just couldn't imagine the fireball on the grainy video display at the Atomic Museum emerging from the casing of the bomb. When I started in applied entomology, I could not reconcile the minuscule volume of insecticides with the scale of their destruction. Try to picture spraying eight ounces of liquid and killing a million grasshoppers. Now I understand the physical principles of nuclear bombs and the biochemistry of toxins. But I still can't reconcile their size with their power.

My father was not prone to emotional discussion of his work, so his impassioned analysis of nuclear strategy made quite an impression when I was fourteen. He was deeply concerned about our development of computer-guided cruise missiles that could place nuclear warheads within a few feet of their target. Being a know-it-all teenager, I suggested that it seemed like a good idea to be able to ensure that bombs landed where you were aiming.

"No," he insisted, "we don't want to be able to hit the target with any accuracy. A strategy of deterrence means that we intend to launch our missiles only in retaliation. Ballistic guidance is all we need." I wasn't sure what ballistic guidance meant other than something to do with the path of a rock. So I tried to interject the "horseshoes and hand grenades" line, but he was serious. "The only reason for pinpoint accuracy is to destroy hardened targets, missile silos. And if we're aiming at missile silos it means that their missiles are still in place. It means we've launched first."

I don't know if he was right; I'm certainly not a military strategist or a political scientist. But I know that his tone of voice meant that this matter was not, for him, a political abstraction; it was a critical challenge to what he did and why he was doing it. He believed that the strategy of mutually assured destruction required only a defensive capacity. My father was willing to participate in the incongruous process of an arms race to prevent a war, but he wanted no part in starting Armageddon. For him, the notion of a limited, preemptive nuclear strike was as hard to imagine as the fire in the belly of Fat Man had been for me.

We now have *precision farming*, the agricultural equivalent of the cruise missile. Using geographic positioning systems and remote sensing via satellite imagery, we deliver the warheads of industrial agriculture—fertilizers and pesticides—with pinpoint accuracy. With the computer models of pest-population dynamics, we don't have to wait until the insects damage the crops, we can strike first. I should know; I've developed satellite imagery, computer models, and targeting systems for rangeland grasshoppers. My younger brother collaborated on the project when he was programming satellite systems for a defense contractor (he has since become an ecologist, but at least he contemplated a career path as despicable as mine). For those who are reticent to have their food saturated with chemicals, we have taken our weapons to the next stage. Now we can wage biological warfare, releasing predators, parasites, and pathogens on putative pests. This is where I finally drew the line in 1989.

The U.S. Department of Agriculture's plan to import pathogens and parasites from Australia and Asia for distribution across North America to permanently and irreversibly suppress grasshoppers was incomprehensible. Fungal spores, like radiation fallout, mustard gas clouds, and pesticide runoff, can't be constrained. And while radiation and chemicals have half-lives that eventually diminish their harm, living organisms have doubling times. The USDA was willing to risk the biodiversity (there are more than four hundred species of grasshoppers) and ecological integrity (grasshoppers play vital roles in cycling nutrients, regulating populations of poisonous plants, and sustaining populations of wildlife) of the nation's grasslands in the name of suppressing a dozen pest species that cause significant damage to the prairie every decade or so. Mutually assured destruction had come to the prairie. To solve the problems of imported "natural enemies" going astray (which are no more "natural" when brought from other continents than a core of uranium is to Hiroshima), we now incorporate the pathogens—or at least the genes for their toxins—directly into the plants. So agriculture now has a single product that brings chemical, biological, and genetic warfare into one neat weapon, which also turns out to be our food. No problem of national and agricultural defense is beyond technology. Of course, understanding how a technological "solution" works is much easier than attempting to comprehend the essence of a problem it is supposed to solve.

My father patiently explained to me not only how an A-bomb worked but the necessary evil of its existence.

The politics made more sense than the physics, probably because my eighth-grade history class confirmed the essential facts that justified our act. The Japanese had started the war in the Pacific by bombing Pearl Harbor —who starts the fight makes a big difference in terms of moral reasoning when you're a kid. Furthermore, the Germans were also trying to develop the bomb—being able to claim that others were doing the same thing is a justification that resonates with a thirteen-year-old. Finally, what if we didn't drop the bombs? The experts asserted that both sides would have lost more lives in an invasion than were lost in Hiroshima and Nagasaki. Even a kid knows that in a lifeboat, you throw a few overboard to save the rest. Dropping the bomb was a regrettable act, but sometimes you do terrible things to avoid even worse outcomes.

As for the insects, the same arguments apply. After all, they started it by transmitting disease and eating our food. To justify annihilating the enemy, the Department of Defense had only to recall squadrons of Japanese Zeros over Hawaii; the Department of Agriculture had only to recollect swarms of Rocky Mountain locusts over Wyoming. Furthermore, humans are just doing what all life forms do: compete to secure resources using the means available. Plants have thorns, snakes have venom, and humans have pesticides. Finally, what if we didn't spray DDT on malarial villages or malathion on West Nile infested cities; what if we didn't stop the locust swarms of Africa or the grasshopper outbreaks of Wyoming; what if we didn't poison the Japanese beetles, the German cockroaches, the Russian wheat aphids, and the Chinese scales? It's easy to win the game of What If? as long as you get to make up the abstract consequences. Coming to terms with concrete reality is sometimes more difficult.

Before a grasshopper-control program, the prairie swarms with thousands of lives engaged in eating, mating, resting, moving, and singing, perhaps not unlike Hiroshima in the summer of 1945. If one returns the year after a treatment campaign, the poison will have dissipated, the scavengers will have cleaned up the dead, the survivors and immigrants will have recolonized the landscape, and a soft buzzing and bustle of life will have recommenced, perhaps not unlike Nagasaki in 1960, when I was born. Immediately after the spraying, however, the insects stagger about, bodies twitch, the corpses lie baking in the sun, and the silence is broken only by the whisper of the wind, perhaps not unlike Hiroshima on August 7, 1945. At least my dad had the courage to give me John Hersey's *Hiroshima*; I haven't taken my kids to see my work immediately after a treatment.

I was perhaps fifteen when I read *Hiroshima*, and I can still vividly recall the mental images of melting faces and sloughing skin from a quarter century ago. But then there were the other images, those that made another atomic bombing imaginable: the photojournalistic depictions of the Soviet Union as a bleak place, where people were stripped of their hope and earnings. The slogan "Better dead than Red" made sense given the utter destitution of Soviet life. Our politicians created vivid images of red armies and yellow hordes. In the face of such a powerful and dangerous enemy, the need for having—and perhaps using—nuclear missiles to protect our values and property was apparent to anyone, even a kid.

I didn't read *Silent Spring* until I was in graduate school; the heaps of dead birds and the rafts of dead fish were stark reminders of what poisons have done. However, these images were up against stiff competition from my childhood. The August 1969 issue of *National Geographic* was unforgettable. The rich language (the article was titled, "Locusts: 'Teeth of the Wind'") and stunning photos of a courageous locust-control campaign conjured up visions of "tactical briefings," wall maps depicted the sweep of clashing armies across the continent of Africa, and descriptions of "fighter command flying daily sorties" invoked a sense of heroic struggle. I learned the power of images and have a series of photos of grasshopper outbreaks from around Wyoming showing grasses reduced to stubble, sagebrush as woody skeletons, and yucca as shredded tassels. This is the world the enemy leaves us, stripped of hope and earnings. In the face of such a powerful and dangerous enemy, the need for having and using poisons to protect our valued crops and property should be evident to everyone.

For some people numbers reflect reality, and I understood the quantifiable arguments for nuclear weapons: A marine invasion of Japan would have cost a million lives; the Soviet Union was more than twice the size of the United States; Chinese outnumbered Americans four to one. And I keep my own numbers close at hand: During an outbreak there are more grasshoppers in an acre (the area of a football field) than there are humans in Los Angeles; during the 1998 outbreak in the western United States, there were 3,500 grasshoppers for every man, woman, and child on Earth, and they ate twenty million tons of forage; one-third of the world's food production is lost to pests. Images versus numbers, feelings versus thoughts: Scientists operate best in quantitative terms. To the rational mind engaged in developing weapons of mass destruction, morality becomes a quantitative exercise to determine the lesser of evils.

The United States and Russia were superpowers, battling for world supremacy. I was proud to be the son of a man engaged in the testing and development of nuclear weapons. The heroes of the Cold War were not brave soldiers like Sergeant York and Audie Murphy; they were the scientists in the trenches, or to be more precise the tunnels, of Nevada. This war was being waged in dark, underworld labyrinths surrounding a fireball. It was in this hell, this place of necessary evil, that democracy would hold back totalitarianism, where capitalism would prevail over communism. Science and technology had become key to winning a modern war. We'd won the race to put a man on the moon, and it was only a matter of time until American ingenuity would win the race to establish democracy and capitalism across the Earth. However, underneath such grand abstractions was a very personal and concrete fear.

It was often acknowledged that Albuquerque would be a prime target of the Soviet missiles. Sandia National Laboratories, embedded within Kirkland Air Force Base, was a strategically vital center, and my house was a few blocks from the base gates. We moved to the edge of the city in 1970, but I didn't figure that this did much to improve the likelihood of surviving an attack. And so, it was reasonable for a kid to wonder what the Russians had against me and whether (and why) my country really had any interest in taking over the world. I certainly had no interest in being vaporized or in moving to Moscow. In trying to explain the mess we had created, my father never pretended the issue was as simple as the politicians proposed. He explained that the Russian people had no interest in harming me, but they did not control their government's actions. And no, our country did not want to take over anybody; we were simply defending ourselves and our way of life. I suspected that a father in Leningrad would have offered his son the same explanations about defending their way of life and not having any interest in invading Chicago. But in an increasingly crowded, mobile, and interconnected world, isolationism was a fantasy. Even then, the world had become too small and interdependent for us to go our own ways. Peaceful coexistence of mutually exclusive ideologies simply was not an option. But when studying a display at the Atomic Museum and estimating where your house would be if a nuclear bomb was dropped on your father's workplace, it doesn't help much that the enemy was a thought, not a person, and the spoils of war were lifestyles, not land.

The war that is being played out in agriculture is a contest between humans and the natural world to defend our "way of life"—our consumption patterns, our economy, our expectations of comfort, and our demands for convenience. The enemy is the ideology of nature—the notions that resources are finite, that the future has value, that other species have inherent worth, and that the planet is sacred. The grasshopper is not the rancher's enemy any more than the Muscovite was my father's enemy. The insects are a manifestation of a conceptual limit, a demand for shared resources, a constraint on production. We use pesticides as a necessary evil to defend ourselves, our "way of life." To the insects, humans are pests and industrial agriculture is the manifestation of an inimical system of governance and economics. But grasshoppers don't have such lofty thoughts; they don't have enmity toward humans. With utter indifference, they simply pursue their lives as they know how. But with human influences now reaching into virtually every habitat on Earth, we are too deeply entangled in global ecosystems to call a truce. The distribution of the human population and our effect on the planet makes a peaceful coexistence of natural and industrial ideologies impossible. While the battle rages at a global level, I struggle with personal paradox. I develop methods for killing millions of grasshoppers, but I gently capture insects that become trapped in my house and release them outdoors. When the killing becomes direct, intimate, and individualized, the abstract enemy gives way to the beauty and complexity of a single life.

In 1997 I accepted a Russian graduate student into my research program. On our first day together, I was telling Alexandre about Robert Frost, my favorite American poet. He smiled and recited Frost's "Fire and Ice" from memory. When Alex uttered the lines "I think I know enough of hate / To know that for destruction ice / Is also great / And would suffice," we both understood that the Cold War sufficed to destroy the Soviet people. Although Frost wrote the poem forty-five years before the Atomic Age, we had grown up with the choice between the fire of a nuclear holocaust and the icy cold of economic collapse. But we also knew that our generation had inherited the task of building a world out of the socioeconomic rubble. In the past few years, Alex and I have stood together next to Minuteman missile silos outside of Cheyenne and walked the shady riverfront in the formerly closed city of Irkutsk. There's something defiant, even a bit angry, in these small acts, as if we wish to purge ourselves of a conflict that we didn't start and have no interest in sustaining. Over beers with Scott, my research associate, we discovered that he and Alexandre had both served as artillery officers in their respective armies. Each knew how to launch battlefield nuclear shells the size of my kitchen trash can at the other. These are two of the finest men I know; they live just eight blocks apart.

I enthusiastically supported Alex's application for permanent residency. The Immigration and Naturalization Service determined that his skills and knowledge in grasshopper biology, ecology, and management based on years of research in the All Russian Institute of Plant Protection met the criterion of fulfilling a national need. Now he, his wife, and two daughters can stay in the United States. This was a hard decision for Alex, as he deeply loves the Russian culture, but the damage inflicted by the Cold War virtually eliminated opportunities for his family. I suspect that my father is proud of my efforts on Alex's behalf. He might have considered the Soviet system to be the enemy, but he would have liked nothing better than befriending a Russian family

I've come to know and respect grasshoppers and ranchers; both are doing their best to survive. They are embedded in a war they did not start, and they are both suffering. After fifteen years of working as a double agent, I see that grasshoppers and ranchers make utterly absurd soldiers in the war that pits human greed and supremacy against the needs and limits of nature. Neither has any compelling interest in harming the other, but they have been unwittingly dragged into a potentially lethal conflict by forces that neither fully understands. The rancher knows that economic survival requires that his cattle graze right at, and often beyond, the edge of sustainability, he knows that environmental regulations are tightening the conditions under which grazing is allowed, and he knows that the grasshoppers take forage that his cattle need. The grasshoppers know that the grass and broad-leafed plants sustain their lives, they know how to survive droughts and blizzards, and they know that they are part of a community that has emerged and adapted to change for ten thousand years.

I know that with over one hundred species of grasshoppers in Wyoming, with their phenomenal reproductive potential, and with the ability to match their needs to the productivity of the land, the rancher will lose under the rules of industrial agriculture.

My dad was on the winning side. And if good ultimately triumphs over evil—a faith that sustains modern religion and culture—then victory is the final measure of moral superiority. The collapse of the Soviet Union, the Evil Empire of the late twentieth century, was not just a political and economic event; it was a quasi-religious validation, a cosmic endorsement of American values. What appeared to be an insane strategy of feeding the military-industrial complex massive resources eventually won a war of attrition against tyranny. We need no monument on the Washington Mall to celebrate our victory in the Cold War; there is a McDonald's in Moscow and an elected Russian president. What more could we want?

If history suggests that my dad was right, that peace through strength is successful, that capitalism trumps communism, that democracy overrules tyranny, then what of our war against nature? What about the erosion of soil, the contamination of water, the thinning of the ozone, the warming of the planet, the disappearance of species, the decline of fisheries, the death of forests, and the loss of the family farm? In particular, what of my battle—the effort to develop strategies to sustain a human presence on the rangelands of Wyoming? Can we deny the failure of an agricultural system that destroys land and people? Can there really be any question that we're losing? If we use the standard that material victory implies moral righteousness, then our economy of consumption, our politics of possession, and our lives of acquisition are profoundly misguided. If the success of the American lifestyle was tentatively affirmed by the result of the Cold War against the Soviets, then its ultimate failing is as surely demonstrated by the emerging outcome of our war against nature.

In the end, the Soviet Union could not sustain the flow of resources to its weapon systems, armed forces, and far-flung allies in the face of our massive economic investments in these same areas. By applying constant pressure on the Soviets to develop new weapons and more troops at home while attempting to sustain revolutions, subsidize guerrilla wars, and prop up allies abroad, we wore down their system, exposed its weak links, and watched it crumble. Now we are on the other side—attempting to sustain a flow of resources to support industrial agriculture, subsidize the timber industry, prop up the beachfront developers. In agriculture, the search for better weapons is frenetic—developing fertilizers to battle the enemy of erosion, importing predators to attack native species that feed on our introduced monocultures, and genetically modifying crops to overcome climate change, herbicides, and salinization.

In my field, the investment of resources in this war of attrition is waning, the weak links are snapping, and the system is crumbling. When I started in grasshopper pest management, there were three insecticides approved for rangeland. During the 1990s there were just two chemicals in our arsenal, but a new registration last year allowed the arms race to break even. A similar story can be told in nations that battle locusts. Despite having applied more than a million gallons of malathion to the rangeland in 1985-1986 for grasshopper control, Wyoming lost thousands of tons of forage the next year. The impossibility of winning this battle was effectively admitted when the USDA canceled its rangeland grasshopper survey and treatment subsidy programs in 1997. At about the same time, the United Nations' Food and Agriculture Organization made a similar decision in Africa. While the USDA was closing its Rangeland Insects Laboratory in Bozeman, Montana, international laboratories devoted to locust control were being downsized in England, France, and Germany.

With the dismantling of survey programs, grasshopper and locust outbreaks developed in North America, Africa, and Asia almost without warning. In 1998 across the western United States, an area the size of New York State was heavily infested and went untreated. The total rangeland grasshopper population outweighed the human population of the seventeen Western states; the loss of forage was estimated at $400 million. That year's locust outbreak in Madagascar is finally subsiding, but the losses are not yet fully known. And the locust plague in Kazakhstan covered 35,000 square miles at that time and spread into Uzbekistan and Russia before receding last year. Loans and aid flowed into the region, but the infrastructure had so deteriorated since the collapse of the USSR that communications, roads, and equipment virtually precluded effective management. In the western United States, the infrastructure for grasshopper control is also crumbling. Few people know how to manage a large-scale control program, the equipment is inadequate, the chemicals are expensive, the labor force is depleted, the logistics are intractable, and the ranchers are told to fend for themselves. They have a lot in common with the Russian graziers that I met: Both are struggling to retain their dignity and sustain their way of life, while the lies of their leaders and the falsehoods of their systems are inevitably played out.

Many people feared that the Cold War would conclude with an incinerating flash, but the end has been a gradual, painful, complex collapse of Soviet society. Rather than an instant of agony, the suffering is prolonged as vital resources shrivel, essential services wither, and crime flourishes. We have shaped our myths and stories around heroic victories, dramatic transitions, spectacular transformations, and stunning catastrophes. Like the books and movies that depicted a post-nuclear holocaust world, the environmentalists have forecast the imminent demise of ecosystems, tidal waves, global famines, pandemic cancers, and genetic disasters. Paul Ehrlich's *The Population Bomb* managed to combine the images of war with those of environmental disaster. But like the ICBMs, this "population bomb" never fell; the dire predictions faded into a muddled, ill-defined degeneration of human and natural communities. This is how modern wars are likely to end. The war to dominate nature on the Wyoming ranches, the Midwestern farms, the Southern forests, and the Eastern lakes is being lost incrementally. The quality of human life and the well-being of the land are eroding almost imperceptibly. Within a single generation the deterioration is nearly indiscernible. We wait for the bang while listening to the whimper.

For the most part, family ranches, small farms, and rural communities are fading away, as technologically sophisticated, capital-intensive "solutions" to the problems of industrial production flow from our national laboratories and state universities. To the modern agriculturalist, the small farmer is a traitor, the family ranch is a camp of the enemy, and rural communities undermine the "big is better" philosophy. Taking a page from the Cold War, corporate agriculture joins forces with politics to wage a war of economic attrition. Abandoning programs that might sustain farm and ranch communities, federal agencies subsidize the agricultural-industrial complex for the costs of fossil fuels, groundwater, pesticide development, and genetically modified seed. But the defeat of our culture's rural manifestation of nature's ideology is a token victory, the equivalent of an East German Olympic medal. We know all too well that the real war is going badly, despite having poured enough resources into some battles to forestall defeat. As the physical, emotional, and spiritual supply lines necessary to sustain a dying ideology are depleted, we desperately seek to salvage our dignity. Like the Soviets who managed to delay their inevitable collapse through Strategic Arms Limitation Talks, we prolong industrial agriculture by seeking treaties with nature, such as the Endangered Species Act, the Clean Water Act, and the National Environmental Policy Act.

So what do should we do? Continuing our courageous and proud efforts to win a war against nature appears utterly doomed. Unilateral disarmament seems impossible. For culture to survive with any modicum of its present virtues, we will need to continue changing landscapes, managing resources, and shaping ecosystems. Perhaps a gradual and sustained arms reduction is plausible. If Pandora's box can't be closed, at least we can have a world where atomic bombs and pesticides no longer define the terms of conflict. Slowing the rate of collapse makes sense only if we use the time wisely. The walls are crumbling, but it is not too late to negotiate the terms of surrender. Maybe we should learn from the Soviets' mistakes and prepare to lose gracefully, creatively, and humanely. We should be sure to salvage the best—the great books, the finest artworks, the viable communities, the closest friends, the intact grasslands, the old growth forests, and the healthiest farms.

The people who suffer the most in defeat are those most heavily invested in the losing system, the Soviet pensioner and the Oregon logger. Those who suffer the least, the Russian entrepreneur and the Vermont organic farmer, are prepared to adapt to the conditions of the victor. The professor in Novosibirsk and the rancher in Fort Laramie are on the cusp. My father's job as a physicist was to contribute his talents to prepare America to beat the Soviets; my job as an entomologist is to prepare us to lose to nature. We both work from within the system, believing that change is gradual and arises from patience and persistence. If there is a nuclear holocaust or a global environmental catastrophe, then I suppose we must each share in the blame. But if we can contribute to moving human society through this phase of self-destruction, if my father's grandchildren can be spellbound by the beauty of St. Petersburg and enchanted by the biodiversity of the Wyoming prairie, then perhaps we'll each deserve some credit.

My dad doesn't work with nuclear bombs anymore, but he's still involved in underground explosions. Rather than detecting the fragments of colliding atoms under the Nevada desert, he is developing ways of detecting buried land mines. The Cold War may be over, but cleaning up the mess we left behind has barely begun. I still work with insecticides, trying to figure out ways to use less poison while keeping good stewards on the land. I know that even if we manage to transform ranching gracefully into a sustainable system bounded by the ultimate constraints of nature's ideology, there will still be quite a mess to clean up. Endangered species, toxic wastes, depleted soils, and overgrazed pastures are the land mines of my war.

A year ago, my dad called me to consult about a termite infestation he had found under the laundry room. Thirty years ago, he would probably have dusted the area with chlordane. But now he was very concerned about the toxic effects of the treatments the exterminator had proposed. I looked up the chemical and found it was a member of the class of compounds I am working with as part of my research. The chemicals are growth regulators that interfere with the hardening of the insect's exoskeleton after a molt. They have extremely low toxicity to mammals, birds, and fish. I assured my dad that the danger was minimal and the potential damage to the house was worth the risk. At least I could explain that these insecticides were much safer than anything we had when I was a kid.

When I was growing up, my parents never demanded perfection and never accepted mediocrity. They would have replaced the ethical admonition "First do no harm" (*Non noli nocere*)—a standard popular with environmental and medical fields—with the more realistic principle admonition "First do some good." We were a family of positive incrementalists, wherein the task of life was to constantly do better, one step at a time. If political systems and ecosystems erode through gradual, nearly imperceptible decay, then the same dynamic will be the process through which they are restored and sustained.

My dad and I can't claim that we've done no harm in our careers. But a world with land mines is better than a world with H-bombs, and a world with insect growth regulators is better than a world with organochlorines. He didn't revolutionize physics, and I am not revolutionizing entomology. We won't leave the world a perfect place at peace with itself, but I like to think that a rancher outside of Cheyenne will almost certainly have a lower chance of being caught in a cloud of radioactive fallout or poisonous spray, and that's worth something.

In the final analysis, maybe he made a mistake in working with nuclear weapons, and perhaps I am wrong to be working with pesticides. Our critics would perhaps legitimately claim that he perpetuated a potentially deadly strategy that has not yet been fully vindicated and that I sustain a lethal strategy that is ultimately doomed. He is still working to prepare the world for what it means to have won, and I am laboring to position us for what it means to lose. I don't know if we're right, but I know that when I was ten, there were 39,000 nuclear warheads in the United States and the USSR; just before my daughter was born, there were 70,000; and when she turned twelve, there were about 10,000. I know that in 1987, we blanketed a typical 10,000-acre grasshopper infestation with five tons of a neurotoxic insecticide, and this year we used forty pounds of an insect growth regulator, applied to just one-third of the infested land. I don't know if doing less evil is the same as doing good, but it's better than doing nothing. I don't know if gradual, continual progress from within our roles as bit players in the military-industrial complex and industrial agriculture will be sufficient to create a healthy human community embedded within a vibrant diversity of ecosystems, but then I don't know what else to do.

Confessions of a Deserter

*Science is the art of creating suitable illusions which
the fool believes or argues against, but the wise man enjoys their
beauty and their ingenuity, without being blind to the fact that
they are human veils and curtains concealing the abysmal dark-
ness of the unknowable.*
—Carl Jung

IN THE CULTURAL war between science and religion,
I would like to be a conscientious objector. However, in
graduate school, devotion to science was virtually a condition of future employment. I joined academia and attained the rank of full professor by teaching, publishing,
grant writing, and otherwise proving my loyalty. After
being a scientist for this long, I suppose I cannot claim
to be a conscientious objector. And so I confess to being
a deserter.

As I understand the conflict, my general is E. O. Wilson, Harvard University professor, two-time Pulitzer
Prize winner, acclaimed sociobiologist and conserva-
tionist, and de facto spokesman for the life sciences. For

my part, I am a pretty good scientist, with a decent rep-
utation for knowing a whole bunch about grasshoppers,
attained while working at a not-bad university. He has a
chest of medals and ribbons; I have a few stripes. Like a
good general addressing the troops, Wilson enthusiastically predicts the triumph of scientific materialism (the
contemporary form of logical positivism) in his recent
book, *Consilience: The Unity of Knowledge,* which might
have been better titled *Conquest.* He eagerly anticipates
conquering the arts, capturing the humanities, and oc-
cupying religion under the aegis of evolutionary reductionism. But if I am commanded to take up arms against
the human spirit, I appeal to the Uniform Code of Military Justice, which allows soldiers to defy an illegal
order. I appeal to the greatest virtue of science in the
modern world, a radical egalitarianism that transferred
the "power to know the truth" from the whim of authority (whether political, religious, or, ironically, scientific) to the shared, firsthand experience of the people.
Contrary to Wilson's interpretation of history, the in-
heritors of the Enlightenment were people like Ralph
Waldo Emerson, not Friedrich Wilhelm Nietzsche.

When I officiated high-school basketball to earn a bit
of money in college, there was a saying that the differ-

ence between high-school, college, and professional ref-
erees was that in high-school games the ref says "I call
them as I see them," in college it's "I call them what
they are," and according to the pro, "They are what I
call them." Taxonomy, the foundation of all biology, is
very much the same as basketball officiating. The expert
who has legitimately earned the respect and acquies-
cence of the scientific community through devotion,
hard work, consistency, and experience is granted the
power to arbitrate the classification and naming of life
forms. But in science, the power to define reality is subject to revocation via peer review. Moreover, one's authority is transient: Future scientists can overturn
previous decisions. As such, science is a living creation,
operating through the tacit acceptance of society. Com-
pared with royal proclamation or papal edict, science's
reliance on earned authority and peer consensus is a reasonable and legitimate approach to organizing human
understanding of the natural world. But it can be unsettling to experience the "they-are-what-I-call-them" creation of reality through this subjective process.

In one of my early submissions for publication in a
scientific journal, I was taken to task for using the term
grasshopper *communities*. A reviewer asserted that I was

clearly a neophyte in the field and that there was no ecological evidence that grasshoppers formed communities
(systems of organisms with a high degree of interdependence). I took my lumps, replaced *community* with
assemblage (an ecologically neutral term), and had the
paper accepted.

A couple of years later, I was preparing a manuscript
on the extinct Rocky Mountain locust, the insect that
blackened the skies of the early pioneers and then disap-
peared at the turn of the last century. We had found
some fragments of grasshoppers in the glaciers of Montana and determined them to be the remains of the
Rocky Mountain locust. However, identifications based
on incomplete specimens are dicey, so I enlisted a colleague, Dr. Robert Pfadt, to take a look at the material.
This was a strategically brilliant maneuver, as a reviewer
of the paper even went so far as to comment that while
my expertise in the area was questionable, if Dr. Pfadt's
forty years of work on grasshoppers decreed that the fragments were those of the Rocky Mountain locust, then
the evidence was incontrovertible.

Now that I have more experience in grasshopper ecology, biology, and management, I can make my own

claims. My experience has convinced me that grasshoppers form communities, functionally cooperating
and facilitating one another's existence in a myriad of
complex relationships. What had changed was not the
grasshoppers, but my perception of the grasshoppers
and the scientific community's perception of me.

In a complex, information-rich world, empowering a
class of learned elites to define reality has some virtues;
the hazard comes when we begin to think that scientists have exclusive access to absolute truth. I don't
want freshman biology students renaming species, but
neither do I want them dumbly acceding the essence of
nature to scientific authority. While scientists might
endlessly frustrate the public with extensively qualified,
tenuous conclusions (e.g., "It appears that under the
conditions in which our observations were made, this
chemical may have the potential to. . ."), they are sin-
gularly unwilling to allow other epistemologies, or ways
of knowing, to share the stage. Science has an immense
stake in retaining and expanding its power, as such authority translates into tremendous social resources. For
E. O. Wilson, perhaps the most strident scientific ma-
terialist of our times, morality can be empirically de-

rived and religion can be dismissed as an evolutionary by-product.

I would not advocate a form of epistemological relativism in which any question can be answered with any method (mathematical theorems are not proven by prayer, despite my desperate entreaties during geometry exams), but for the big questions of the world, there needs to be scientific humility and epistemological pluralism. The question of human existence is a matter of both how we came to be (the legitimate purview of science) and why we came to be (the valid realm of religion). For scientific texts to assert that there is no purpose or direction in evolution is as unfounded as the religious tracts that claim to interpret fossil evidence and geomorphology in terms of the Bible. Science has no method, test, or instrument to detect meaning in the universe.

Scientific humility is hard to find when the stakes for cultural conquest are immense and growing ever larger. In a society that fills the days of its people with the complex process of acquiring and consuming a growing list of goods and services, we are coming to the point where we are too busy to figure out for ourselves what is real. So we fund our sacred and secular authorities to work it out for us, dividing our faith and our dollars between re-ligion and science in the hope that the Invisible Hand of God and the Mystery of Mathematics will resolve the big questions without our having to participate. In such a competition it is no wonder that science claims epis-temological superiority. The spoils are not just enor-mous wealth and prestige, they literally include the human mind.

When my son was about three years old, he had the usual confusions about the rules of language. He also exhibited what was either an interesting grammatical error or a compelling metaphysical insight. Ethan used the pronoun *who* for all objects in his world. He spoke of the "blanket who I slept with" and the "tree who had the bird nest." Both he and his sister are very empa-thetic and gentle children, and Ethan's language seemed to express his belief that he was immersed in a world of fellow beings. Eventually, we played our role as parents and definers of reality and taught him to use *that* and *which* in reference to objects. Maybe someday Ethan will read Martin Buber's *I and Thou* and decide that the world is composed of *who* (or *Thou*) after all.

Buber contends that humans enter into relationships with the world using predominantly an I-It association, precluding the possibility of mutuality or reciprocity. An *It* can be experienced and used, but a person comes fully into being through a *Thou*. Such reciprocity and association "makes life heavier but heavy with meaning." Buber maintains that there are times and places where a relationship cannot be I-Thou by virtue of its goal. A therapist is a patient's instrument to regain health, and an animal may be a scientist's means to an end. In pursuing some legitimate goals, we treat portions of the world as objects. But the human potential cannot be reached by pursuing only the knowledge that comes in this manner. Science is a celebration of the I-It relationship, but as Loren Eiseley asked, "Is there something here we fear to face, except when clothed in safely sterilized professional speech? Have we grown reluctant in this age of power to admit mystery and beauty into our thoughts or to learn where power ceases?"

How much power will science lose if scientists admit the limits of objectivity and materialism? The answer lies, perhaps, in whether science can both maintain its integrity and find a way to allow transcendence. Brian Swimme has proposed such a method by suggesting that what we hold to be objective perceptions of the world are co-creations, emergent realities subject to scientific exploration and interpretation. In *The Universe Is a Green Dragon*, he asserts, "When you stand in the presence of the moon, you become a new creation. . . . the human awareness could never know the throbbing presence of the moon and all the intensity of feelings were it not for the moon itself. These feelings are as much the creation of the moon as they are of the human." This does not portend the end of astronomy, only a deeper understanding of the intimacy between the astronomer and the heavens. As he notes with Thomas Berry in *The Universe Story*, we alone do not create the feeling of awe that a climber might experience on a mountain and the sense of wonder that may permeate the geologist who comprehends the forces and time that gave rise to the mountain. These feelings are neither our subjective fantasies nor the mountain's objective experiences; they are personally significant and mutually evoked. They transcend scientific analysis, but they do not negate its value. As such, science is a necessary but not sufficient mode of understanding. The exploration of meaning complements, but does not reduce, the work of science.

In the 1980s I began my work in applied ecology of rangeland grasshoppers (my favorite euphemism for pest management) by asking how we could most efficiently control their population outbreaks. I perceived these events to be objective, natural phenomena that we could understand and then manipulate. I soon discovered, however, that an outbreak was an event with a reality arising from a remarkable admixture of history (the number of grasshoppers capable of damaging wheat decades earlier had been institutionally fixed as an out-break threshold), policy (the U.S. Department of Agriculture's involvement in grasshopper control defined the methods of actually delineating an outbreak), economics (state and federal subsidies for control programs dictated the minimum area of an outbreak), and science (including the statistical carrying capacity of the rangeland, an arcane calculation with uncertain meaning).

Grasshopper populations declined in the early 1990s, and I struggled to find infestations on which to apply my objective, scientifically derived control tactics. However, as I spent more time with grasshoppers, I came to ask how humans and grasshoppers could exist in a relationship that was co-created by their existence and our perceptions. With this understanding came a remarkable series of ecological events.

The conditions necessary to explore co-existence were abundantly available, as if the grasshoppers were participating in the effort to find a way for ranchers and insects to cohabit the Wyoming prairies. Rather than attempting to totally eliminate large-scale, high-density outbreaks, we now try to thin infestations, understanding the importance of leaving enough grasshoppers in the field to sustain the predator and parasite populations that will reestablish natural constraints. We've developed methods that use 99 percent less insecticide than the tactics available in 1990. Describing these approaches as a co-creation of the grasshoppers and myself would rankle my scientific colleagues and my environmentalist friends. The skeptics of our approach, which involves applying low rates of safer insecticides in widely spaced swaths (rather than high rates of broad-spectrum insecticides in blanket coverage), want to know exactly how and why such a method works before endorsing its use. We are working to untangle the complex processes that account for its success, including the roles of cannibalism, grasshopper movement, and conservation of natural enemies. But for now my best answer is that it works because the grasshoppers, as collaborators in creating the reality of the rangeland, allow it to work.

James Gibson coined the noun *affordance* to describe the mutuality that emerges in the context of the human-nature relationship. For example, in mountaineering a human does not climb a rock; the boulder affords footholds and handholds, creating a passage. In the same way, the scientist is afforded access to the universe not by forcing his or her way through a passive and inert matrix of material existence, but via the paths and openings (affordances) that arise in relationship. Understanding the world through dominance, like forcing your way up a rock face, might provide short-term success, but it feeds a false sense of power. Consider the entomologists who, in the early, giddy days of DDT, told children to be sure to collect specimens of pest insects because insecticides would soon eliminate these species from the landscape. Today DDT is banned and the pests are flourishing.

In a 1998 lecture to the university community, I explained some of my struggles with killing insects while trying to maintain both my scientific credibility and my spiritual accountability. I began with these words:

> I love irony and paradox—they are the source of much creative energy and can become doorways to deeper understanding. I recently encountered Theodore Roszak's argument that modern society, through the collusion of science and religion,

with the passive assent of humanities and the arts, has emerged as the most idolatrous culture in human history. This is a disturbing analysis, given our pride in having—supposedly—driven such misguided perceptions from our worldview. But Roszak seems to be onto something with the definition of idolatry as "a state of consciousness in which the particular is mistaken for the ultimate." In more familiar terms, an idol is a thing or place that is misunderstood to be nothing more than its material existence. For example, one might mistake a grotto, a tree, or a grasshopper for literally being a god, a truth, or an experience, with nothing more lying deeper or hidden—a symbol devoid of transcendence. This sort of ontological error is most commonly associated with pagans, who, it seems, never actually believed the sorts of things that were (and are) ascribed to them. Rather, they perceive objects and places as local manifestations of a universal presence. The supposed idol actually functions as a path to transcendence.

During the questioning, I responded to a student's query about science education with the ironic observation that in biology, the putative science of life, we spend an inordinate amount of time killing things and studying dead specimens. A colleague in zoology jumped in, claiming, "The difference between zoology and agriculture is that we kill to understand. You kill just to kill." I replied that killing a rat and treating the animal merely as an object for dissection is no more noble than spraying a field of grasshoppers and treating them as an obstacle to economic prosperity. One's relationship with another life form was as critical as one's abstraction of social benefit. To which another colleague argued that he killed fish as a means of understanding their physiology, which would eventually translate into improved fisheries management. The central importance of mindfulness and intentionality faded into the background as we discussed whose science had the loftier goals. What I truly regret was not expressing my core understanding: I had entered the land and knew—not through economic rationalization, environmental quantification, or ecological experimentation, but through experience, intuition, and humility—that it would accept ranchers if the grasshoppers were respected and included in the co-creation of the prairie.

To adapt the line of thought advanced by Swimme and Berry, in a biologically meaningful sense, the prairie is a creation of the grasshopper. These insects have sculpted plant communities, cycled nutrients, hosted parasites, fed wildlife, and distributed fungal spores. Upon colonizing the grasslands of North America as the ice retreated, the grasshoppers co-evolved with the soils, plants, and other animals. Understanding the challenges of bitterly cold winters and blistering summers, they committed to an unseen vision, a way of life that drew them forward perhaps through the sheer power of their reproductive potential (grasshopper outbreaks in the western United States often outnumber of the human population on Earth by several thousand-fold) or the staggering diversity of their forms (there are 112 species in Wyoming alone, more than the species of resident mammals, birds, fish, reptiles, or amphibians). The grasshoppers became a whole region of fecundity and diversity, a creation of the entire Great Plains.

Using the gifts of science, we too must find ways to become native to this place. On a planet with nations having staked out every piece of land, there are no frontiers, nowhere else to go. I have listened to the land, and I know that the prairie is mute when we kill too many grasshoppers, that some must be there to sustain the soft breathing of the grasslands. I have spoken aloud to the grasshoppers before and during a summer's work, asked their permission, explained my intentions, apologized for my mistakes, and listened for their reply. They do not resent my work; they accept that birds, ants, and humans will thin their numbers. But their ten thousand years of evolutionary wisdom in this place demands our respect. We need to listen. I wanted to ask my colleagues if they had ever taken the time to ask how the rats felt about biology labs and whether the streams wanted to be fisheries.

I can imagine my colleagues' response to my assertion that the grasshoppers work with me in developing pest-management tactics. It would be much like Hui Tzu's response to Chuang Tzu in the Taoist story about knowledge. During a stroll over a bridge on the Hao River, Chuang Tzu looked down into the water and said, "Look how the minnows dart hither and thither at will. Such is the pleasure fish enjoy." The skeptical Hui Tzu replied, "You are not a fish. How do you know what gives pleasure to a fish?" To which Chuang Tzu answered, "You are not I. How do you know I don't know what gives pleasure to a fish?"

Expressing this notion of transcendent knowledge in more familiar, Christian terms would suggest that such insight arises from prayer. I once asked Joe Fortier, a top-notch doctoral student in our entomology program and a Jesuit priest (now teaching at Gonzaga State University), how he integrated prayer and science. Joe suggested that there was deep complementarity if we think of prayer as "the intentional communication with the divine." If we use William James's definition of the divine as being that which is most primal, enveloping, and deeply true, then I would contend that good science must be prayerful. The key to virtuous science is the discovery of a question or line of research that has authentic value. To ask such meaningful questions, the scientist must consciously and intentionally enter into dialogue with the inescapable essence of being and power. If there is hope for science in this regard, then it is likely to manifest in ecology, the science that claims to have holistic perspective, if not yet attaining spiritual vision.

Although holism would seem to defy reductionism, science has tried to have it both ways. Some, like E. O. Wilson and the physicists alluding to the Theory of Everything, cling to the faith that reductionism will condense the material world (the whole) into a unified principle. But in every field—evolution, cosmology, physics —the more we probe the empirical, the more we encounter mystery. The role of the unknowable, usually labeled "random chance" to create the appearance of formal understanding, is firmly established in the ground of being. In fact, there is no scientific test or proof of randomness; it is simply a phenomenon that we cannot resolve into a predictable pattern. We are coming to understand that the universe is fundamentally intractable and ultimately indeterminate. Our uncertainty is not a function of inadequate experiments, instruments, or equations; mystery is the essence of reality. Scientists once disparagingly referred to religion's God of the Gaps, having driven the divine into what seemed to be ever-narrowing spaces between formulae, models, and data sets. Ironically, it now appears that science has established not only that the gaps are inherent in the universe but that from these unknowable spaces emerge the creative forces of quantum fluctuations, biological mutations, and the origin of the big bang. What science refuses to consider is that having reached the limits of its own epistemology, there remain other ways of accessing mystery.

I co-taught a course on science and religion in which we had the four instructors (a theologian, a sociologist, a zoologist, and me) engage in a panel discussion of holism. From this discussion emerged three conceptual models, best illustrated with an example. Let's assume that we want to develop a management plan for a lake. The "scientific model" of holism would suggest that sound management requires the involvement of biologists, chemists, hydrologists, geologists, atmospheric scientists, and the like. No single scientist could possibly have sufficient expertise in all of these fields. The class generally agreed that scientific holism was a better approach than claiming to understand the lake through any one discipline. However, there was a more expansive approach, the "academic model" of holism. According to this notion, the team studying the lake should be composed not only of these scientists but also of historians, anthropologists, sociologists, political scientists, and economists. This breadth of expertise assured that matters of policy, culture, and valuation were included. In general, the inclusion of the soft social sciences was viewed as an enlightened concession, but this was as far as holism could be taken. The final approach was my "radical model" of holism, in which the team would include a shaman, a poet, and a fish. For the most part, my colleagues and students found this to be absurd—these entities could not even communicate with the scientists. I hesitated to suggest that they might, however, be able to communicate with the lake. In fact, perhaps the shaman and poet were the only beings who might be open to communication with both the fish and the chemist.

We must teach our students and ourselves to ask not only what a lake or a grasshopper or a molecule is and how it works, but also what it means. To do this acknowledges that understanding the nature, origin, and history of something is profoundly important but essentially different from perceiving its importance, meaning, and significance. William James clearly elucidates this difference in *Varieties of Religious Experience*, in which he argues that understanding the psychological origins of spirituality had no bearing on the value of religion. Referring to principles of logic, he shows that there is a critical distinction between existential judgment (what something is) and spiritual/propositional value (what something means). These are fundamentally different issues, neither of which reduces the other. James values religious experience no less for having gained some understanding of its psychological origins.

Unfortunately, this enlightened and clear perspective on the relationship between science and religion has been obscured by biologists, like E. O. Wilson, who advocate the idea that the power and meaning of religion will be destroyed when we gain some understanding of its evolutionary origins. He contends that scientific materialism is the final word on matters of both existential judgment and spiritual/propositional value. As evidence, he notes that moral principles are often consistent with evolutionary constraints, somehow implying that transcendent insight can't be consistent with biologically sustaining practices. In *Biophilia*, Wilson contends that other species warrant our attention because they have instrumental value (they serve human interests). As an evolutionary legacy, our mental health requires frequent contact with other creatures. From this, I would logically conclude that if we were able to shed this baggage with a bit of gene therapy, our obligations to other life forms would disappear. Fortunately, more constructive intellectual perspectives are being offered in other fields.

Rather than ethics collapsing into Wilson's empirically derived, evolutionary imperatives, Susanna Goodin —a philosopher and colleague—suggested in a recent lecture that ethical thinking is expanding beyond traditional Western anthropocentrism. In the most direct form, extensionist ethics applies the principles of anthropocentrism to animals, as commonly seen in the animal rights movement. The biocentrist, however, abandons the notion that humans are superior and argues that all life has a "good of its own" that merits our moral attention. The ecocentrist takes this notion a step further, contending that wholes (habitats, ecosystems, and Earth) have a "good of their own" that must be considered. This expanding sphere of moral concern is seemingly laudable, but it persists in the notion that there is a center around which our reality revolves. Like the cosmologists who debated the heliocentric versus geocentric views, Western philosophers argue about what the center of the moral universe might be. However, the debate may be doomed by the premise that there is a center. Modern cosmology reveals that there is no center, no absolute point of reference. Every point is moving away from every other point, and no special place can be said to represent the origin of the universe. What if our search for unity, our quest for a moral center, our faith in a singular truth, is fundamentally flawed? Suppose we listen to the cosmologists and imagine a polycentric or anthropocosmic ethic, where each being is at a center?

Situational ethics is a way of conceiving of our responsibilities by making love the absolute standard against which all moral decisions are measured. Every conflict must be understood and resolved on its own terms; each relationship becomes its own center. But there is no moral relativism; the only right course of action is to do that which is most loving. Of course, fully understanding both the conditions and the standard is an impossible challenge. For its part, science may reveal more of how a situation came to be and what the consequences of our actions might be, and this knowledge is vital to making the best possible decision. For its part, religion may allow deeper insight to the meaning and purpose of a situation. Our obligation is to do the best we can throughout life, which requires the knowledge of science, the insight of religion, and the humility of realizing that we shall never understand the enormity of facts or the fullness of meaning for any situation.

Physics tells us that we are interconnected in time and space to a web with no center; cosmology instructs us that we are formed of aggregated stardust, energized by the fire of the big bang; ecology informs us that entirely new and unpredictable properties can emerge when individuals interact; evolution reminds us that we are but a twig on the evolutionary bush. These lessons challenge our notion of an autonomous *self*, as do the teachings of religion that refer to selflessness, interdependence, omnipresence, and the unseen. But rather than the complementarity of fact and faith being cause for cultural celebration, it seems to stoke the fires of rivalry. Galileo and Darwin moved the Earth and humans from the material center of creation, but in an ironic turn of history, their works have been used to place science at the center of knowledge. Existential humility somehow led to epistemological arrogance. But if science is the only valid way of knowing and if this methodology shows that all existence is an arbitrary collection of contingent accidents without purpose or direction, then science itself can have no meaning. In this most wicked of ironies, we either dismiss science as one of the meaningless products of evolution (as the reductionist snake grabs its own tail and swallows itself), or we take science seriously, deriving its value from elsewhere.

The essence of the modern problem is that the scientist has claimed the role of a priest (with unique access to truth and power) without accepting the obligations of a minister (with the responsibility for justice and compassion). Parker Palmer cuts to the heart of the issue: "Objectivism, driven by fear, keeps us from forging relationships with the things of the world. Its modus operandi is simple: when we distance ourselves from something, it becomes an object; when it becomes an object, it no longer has life; when it is lifeless, it cannot touch or transform us, so our knowledge of the thing becomes pure." However, he points out that objectivism bred new versions of old evils. The scientist is fond of historical allusions to religious wars, but Mao's Cultural Revolution, Stalin's purges, Hitler's concentration camps, and our Manhattan Project were the products of fiercely objective reasoning. If science, by virtue of its own assumptions and methods, can deduce no purpose, find no cause for hope, or extract no transcendent insight, then the human spirit will seek elsewhere or wither for want of meaning.

I've been admitted to the priesthood of science, although I risk excommunication. Using a quarter of a million dollars in public funds, I have embarked on a three-year mission to refine our methods of waging war on the natural world. In the course of this project, hundreds of millions of grasshoppers will be killed along with untold numbers of other insects. So what of ministry; where is the compassion and justice? Each summer I renew my spiritual struggle by resolving to kill well. I seek to become a loving assassin, deepening my relationship with my victims—and with my accomplices. For like the grasshoppers, ranchers are easy to vilify, have hard exteriors, and are not easily understood. And so my meaning lies in the faith that I can bring a sense of justice and compassion to the world of the grasshoppers and ranchers, for each is engaged in a fierce struggle with an enemy they do not understand.

The Battle of New Orleans was fought on January 8, 1815, fifteen days after negotiations in Belgium had officially ended the War of 1812. For more than a thousand men who died in the bloody fighting, the news of the truce did not arrive in time. At the turn of the nineteenth century, Henry Adams described the history of Western culture as a battle between *civitas Dei* (the order of God, or the Commandments) and *civitas Romae* (the order of Rome, or the Law), a power struggle between spiritual insight and human reason. Throughout the twentieth century, the battles were fought in courtrooms, lecture halls, and pulpits. But as we enter the twenty-first century, one must wonder if the scientific materialists and religious fundamentalists are not the modern equivalents of the British and American troops in 1815. The battles rage on, but the war is over. As the vultures of radical relativism and the jackals of deconstructionism pick over the rotting corpses of the logical positivists and creedal absolutists, the conscientious objectors and deserters have called a truce.

Scientists like me (a Unitarian ecologist) and theologians like Joe Fortier (a Jesuit evolutionist) may not agree on exactly how to interpret the world, but we understand that there is no final victory to be had, no ultimate refutation of religion or eventual overturning of science. The war between science and religion can't be won any more than a paradox can be resolved. Science allowed me to see the grasshoppers; religion gave me the ability to listen to them. To see without listening would make them into mere objects; to listen without seeing would make them into spectral voices. The truths of science are the gateway to spiritual insight, and religious truths give meaning to the work of science. Neither needs the other, but humanity requires both.

The Meaning of One

The definition of the individual was: a multitude of one million divided by one million.
—Arthur Koestler

THE TINY FORM was soon to be laid in a shallow trench in the dank soil—it was the least I could offer to one who shared such an essential quality with me. My gaze drifted over the field to the tree line a half-mile away. According to my hosts, the Vermont spring had been unusually wet, and the field was overgrown in a tangle of grasses, sedges, herbs, and dandelions. Far from home on the Wyoming plains where rain is measured and celebrated in tenths-of-an inch, I observed the verdant Vermont field overflow with life. Although I pride myself in being a decent field ecologist, few of the plants were familiar except, to my embarrassment, the dandelions. I have traveled to fifteen countries across six continents, and perhaps the most common floral associate of civilization is the dandelion. It seems to be nearly everywhere humans are found.

The dandelions in this field were unusually robust and plentiful. The flower heads, now mostly turned into delicate puffs of seed like so many perfect white spheres, stood nearly two feet tall. I've been told that dandelions represent a classic example of natural selection. In a lawn that is frequently mowed, only the shortest dandelions manage to avoid decapitation and spread their seeds, so urban landscapes are dominated by stunted, dwarf dandelions. In uncultivated settings, only the tallest dandelions successfully disperse their seeds into the breezes that waft over the competing canopy of grasses and herbs, so meadows produce big, strapping dandelions. Judging from the towering flower heads, the field in front of me must not have seen a mower more than once or twice a year. At least the little, still one would have a relatively undisturbed setting in which to rest.

In Wyoming, where bluegrass lawns are tributes to human folly and an excess of leisure time, dandelions are the nemesis. Actually, my attempts at weed management are a bit half-hearted, and given my neighbor's steadfast refusal to control his dandelions because the goldfinches eat the seeds, my battle for a monocultural lawn is a lost cause. It's odd that the dandelions in the Vermont pasture didn't seem like weeds. But then, maybe they aren't. A colleague at the university who is a weed scientist—which seems to be an odd, almost oxymoronic juxtaposition of terms—defines a weed as a "plant out of place." This absurdly simple and ultimately anthropocentric concept seems appropriate for a scientific discipline primarily devoted to developing a never-ending array of herbicides. But according to his definition, the dandelions flourishing amidst the tumult of lush growth on this New England hillside were not weeds at all. They were, admittedly, a somewhat dominant feature of the field, but they did not seem to be out of place; they didn't seem to be crowding out other plants or disturbing the homogeneity that we seek in our lawns. I know that they are exotic introductions to this continent, but then so am I. Dandelions and I belonged in that lush meadow, or so it seemed, as did the tiny form that I prepared to cover in a thin layer of rich, black soil.

Being a scientist, I am fascinated by numbers, and the dandelions in the field presented a pleasant problem in quantification. At my feet, I estimated that there were about thirty dandelions per square yard, which meant that there were nearly 150,000 per acre. Given that the field extended about a half mile from the dirt road down to the bottom of the draw and half mile from the fence line where I stood to the tree line, the area encompassed about 160 acres. That meant there were about 23 million dandelions in the pasture. Having plucked a seed head and given it a somewhat self-conscious and entirely frivolous puff that sent the tiny parachutes into the breeze, I took a minute to pluck another and estimated 260 seeds per head. Until then, my calculations had been mental, but the final step demanded that I take pen to palm (no paper being available), and I arrived at an estimate of just over 6 billion dandelion seeds in the field.

As I put my pen back in my pocket, I plucked a wayward seed from my shirt—one of six billion. With so many, how important could one be; how much difference could this life make? I bent down, scraped a shallow trench, buried the seed in the earth, sprinkled on a bit of humus, and gently patted it in place. This one life out of slightly more than 6 billion, had briefly changed the course of one conscious life, brought a moment of pleasure, and provided an opportunity for connection. Perhaps we should aspire to such meaning in our own lives.

Epilogue

A vaporous emulsion of lodgepole pine resin, sage oil, and rotten-egg vapors drifted on the breeze. Ethan and I were on our first fishing trip to the beaver ponds outside of Laramie. "Hit it hard, once above the eye," I told my son. The brook trout's bold orange spots were accentuated against the black mud. It had swallowed the hook, and Ethan knew from my earlier explanation that we couldn't return it to the water. I felt a qualm as Ethan picked up the stone that I had used a few minutes earlier. "Go ahead," I told him, "you caught it. You have to kill it." He knelt at my feet, took a couple of tentative half-swings, and then delivered a weak and desperate blow to the side of the fish, renewing its futile thrashing.

Ethan looked up with the pathetic half grin of a six-year-old struggling with a task beyond his capacity. His wide, unblinking eyes made it clear that he knew the fish was worse off than before. "That's OK," I said, gently taking the rock from him and striking the fish. Stunned, it slumped into the mud. "Ready to fish some more?" I asked, slipping the trout into our creel. Ethan recovered quickly and darted over to the worm bucket.

Acknowledgments

This book is forty years in the making, and to acknowledge the contributions of all who contributed to my understanding of the natural world is, of course, impossible. However, I shall try, knowing that those I mention here were vitally important voices in my story, and many of those I overlook at this moment in my life will continue to whisper so that one day I will hear their truths. I must begin by thanking my parents and siblings for having created a safe and loving, if sometimes competitive and raucous, context for growth. The seeds of science, and biology in particular, were planted deeply by three high-school teachers to whom I owe much:

Beverly Cotton, David Magruder, and Ernest Polansky. In college, the biology faculty at New Mexico Tech cemented my love affair with the life sciences, with Drs. Gil Sanchez, Al Smoake, and especially Tom Turney playing unforgettable roles as mentors. In graduate school, my sage advisor, Dr. Rick Story, gently and quietly pushed open the door to transcendent understanding through science and human experience. At the same time, Rev. Steve Crump pulled open the same door from the spiritual side. Since I came to Wyoming, the Unitarian Universalist Fellowship has been a wellspring of acceptance and discovery, with Karl Branting, Hal Wedel, Dennis Parkhurst, Bill Reiners, Gary Smart, Mike Vercauteren, and Zanna McKay being a few of the people to touch my life. The faculty at the University of Wyoming has been at least tolerant of my views in the best tradition of academic freedom (and responsibility), with genuine support being offered by my colleagues in entomology. Other scientists willing to listen, and even encourage, my unusual perspectives include Frank Howarth, Stephan Halloy, Bill Kemp, and Jay McPherson. Even a few administrators, including Joe Kunsman, Tom Thurow, and Steve Williams, have been profoundly understanding. Students have often been my teachers, or at least the catalysts of my learning, including Chuck Bomar, Larry DeBrey, Joe Fortier, Alex Latchininsky, Carlie Miller, Narisu, Rich Rockwell, Spencer Schell, Douglas Smith, Kirk Van Dyke, and Nina Zitani. In this regard, Scott Schell (who has been my student, teacher, research associate, and friend for ten years) has been a model of loyalty, integrity, and rootedness. The students, staff and faculty of Sterling College (Vermont)—and their WildBranch Writers' Workshop (especially my guru, Ted Gup)—provided the stimulation, time, and setting to lay the foundation of this book, and the support of Jeff Bickart will resonate always. Mary Benard, Chip Blake, and Bob Tarutis were the first editors to believe in my work, and their faith in me (and Kim Leeder's devotion to extracting the best from my writing) made all the difference. Finally and foremost, I offer my deepest gratitude to my wife, daughter, and son. Without their unconditional love, this book and its author would not have come into being.